DODGE VIPER

ON THE ROAD

COMPILED BY
R M CLARKE

ISBN 185520 2034

Booklands Books Ltd.
PO Box 146, Cobham, KT11 1LG
Surrey, England
Printed in Hong Kong

ACKNOWLEDGEMENTS

This is the first in a new series of Brooklands Books devoted to outstanding motor cars which are in current production, and we could think of no better subject to launch our series than the Dodge Viper. This extraordinary car is one of those destined to become a legend from the word 'go'.

Brooklands Books are an archive series for motoring enthusiasts, gathering together in convenient form existing material about interesting cars. They depend for their existence on the generosity and understanding of those who hold the copyright to the material they contain, and we are always pleased to acknowledge the assistance we receive in putting them together. On this occasion our thanks go to Autocar & Motor, Automobile Magazine, Car and Driver, Car South Africa, Motor Trend, Road & Track and R & T Specials.

To those people who thought legislation and high insurance premiums had killed off the Muscle Car in the early 1970's, Dodge's startling Viper came as a revelation in the early 1990s. Indeed, for the Viper's designers to get such a car past the bean-counters at Chrysler Corporation was no insignificant feat, and few people who saw the original concept car can have seriously believed that it would one day become a production reality.

R M Clarke

This book contains a report comparing the Viper with the legendary AC Cobra - and not without reason. The two-seater Viper was designed as a "back-to-basics" model which would evoke all the raw excitement associated with cars like the Cobra. Yet the Viper is no mere retro-car; it incorporates all the technological advances which were available to its designers in the early 1990s, and it is all the better for that.

The Viper's all-aluminium eight-litre V10 engine was developed with technical assistance from Chrysler-owned Lamborghini, and it puts out 400bhp with a staggering 450lb/ft of torque. Driving through a six-speed manual transmission, it powers the Viper to 60mph in 4½ seconds and then on to a top speed of 165mph. And its modern technology means that the Viper does not suffer from the poor brakes and handling so often associated with the great Muscle Cars of the 1960's. Not only does it handle superbly, but according to factory figures it can accelerate from a standstill to 100mph and then be safely braked back to zero, all within 14½ seconds.

For those fortunate enough to own a Viper, and for those who can only dream, this book will be a welcome companion.

James Taylor

CONTENTS

I'm riding through the mountainous terrain around Sedona, Arizona, in the experimental V-10 Dodge Viper. Chrysler Motors President Bob Lutz is at the wheel. Over the roar of the rushing slipstream and side exhaust pipes, he's telling me about the car. Just small talk at this point, though, shouted loudly in my direction, punctuated by curt nods of understanding. We're too busy enjoying the moment to engage in anything more serious.

The 8-liter (488-cu-in.) aluminum V-10 roars like a traffic cop at the water slide to hell. When Lutz drops the accelerator, the Viper can't help but lunge forward—mammoth 335/35R17 rear tires grab the ground in a merciless grip. Lutz likes this car. It's his brainchild. The sentiment shows quite plainly on his face as we rush headlong up the smooth, wide mountain roads. We're going fast, passing everything, but Lutz is skillful.

At this stage of development, the Viper mule's instruments are cosmetic. The speedometer is installed but isn't connected. On the flat, gray dashboard, however, is taped a paper chart. It lists a column of gears, a column of rpm, and a column of equivalent mph. When the president of Chrysler Motors passes the unmarked police car like it's a signpost, we're clocking nearly 135 mph.

Chrysler Vice President of Vehicle Engineering François Castaing told us, "We had been discussing some of the cars of the '60s, the Cobra, and at the end of 1987, just after AMC was merging into Chrysler and the Jeep Truck Engineering was established, we undertook the design of the V-10 engine.

"Maybe one day [we thought], like in the '60s, where the Bizzarinis and the Shelbys of the world were using big gas engines to produce a sports car, maybe this V-10 would be so powerful and inexpensive we could bring that back. So, that was the first conversation."

Chrysler Chairman Lee Iacocca announced May 18 in Los Angeles that his beleaguered car company was *going* to bring it back and produce the awesome Dodge show vehicle. The big-block rear-drive car had been displayed to rapturous reviews at the '89 North American International Auto Show, and if any final push in the decision-making was needed, Lutz said, that show response provided it.

The car at once redefines the sports car genre. As the Viper represents yet another attempt at a successful homegrown 2-seat sports car, it's going to stare the Corvette directly in the eye and dare it to blink. The Viper's performance as we experienced it was phenomenal, but its rough edges were plainly evident at this stage. Castaing and Dodge types are looking for achievable 0-60-mph times of about 3.9 sec, with a top speed of about 190 mph. Is this going to be Chrysler versus giant GM—Da-

'92 DODGE VIPER
THE SERPENT STIRS

by Daniel Charles Ross

vid against Goliath?

"Well . . . yes, and no," Castaing said. "We would hate to see the car classified as an anti-Corvette. There is room in this country for Corvettes, and there's room for Viper. They are a different kind of animal."

The phrase Chrysler types use to define the Viper philosophy is "back to basics." A big-block high-horsepower engine, driving the rear wheels (in this car) through a 6-speed manual transmission, is the cornerstone of the formula. The suspension is a simple but elegant double A-arm arrangement.

"The Viper, because of the weight, the power, and torque of the engine, is equipped with quite wide tires," Castaing said. "For that to be put to proper use, whether you use them in the traction mode or cornering, or a combination of the two, you must have enough sophistication in the suspension so you always keep the tire in the proper position on the ground.

"The fact that we're using double-A wishbones like everyone else is because there's no other way to keep the tires on the ground when you need them on the ground. No mystery there. It starts from the tires, and what they're supposed to do for you and how you attach them to the car, to take full advantage of the good tires that Goodyear or Michelin are going to provide to us.

"For the time being, we're trying to get away without using too much magnesium or aluminum, because as nice-sounding as they are on paper, they have disadvantages. Mostly it's the cost of repair if you bumped into the curb or something like that. And that happens with these kinds of cars."

Planners are leaving the frills on the cutting-room floor. In this regard, if no other, the Viper is anti-Corvette, despite the wry denials. Yes, the car probably will have air conditioning, but the windows in the final production car may yet be plastic curtains. Lutz says the top cover isn't even the expected rigid roof panel, but a hard plastic clamshell merely sufficient to keep out inclement weather.

The vehicle we experienced is one of the last design steps to the production model. The changes from the show car are small, an extraordinary thing in view of the Viper's radical look. Actually getting the Viper from show car to showroom is less daunting than the physical deadline of the time frame. Vipers are scheduled to be in production by January 1992.

"The only issue is that we're trying to do it *fast* for a little money. I don't want to say it's *easy*, but it's no major task for a group of smart engineers. It's just doing it fast, so you can produce the car before the end of next year. That much is a challenge."

Chrysler engineers face another challenge as well—handling. "Years ago, Shelby would talk of the Cobra in terms of how fast it could go from zero to a hundred to zero again," Chrysler Vice President of Product Design Tom Gale said. "I don't know how many times that's come up in the good-old-boy discussions Carroll would have.

"And was that an inspiration? *Yeah!* That *was* an inspiration. That was part of the brief for the car. Only this time around, we wanted to have the kind of handling capability that really wouldn't have been there a generation past."

TOM GALE ON THE VIPER

"In February 1988, Bob Lutz had discussions with Carroll Shelby, just one enthusiast to another. Carroll had been wanting to do a sports car for a long time, and we had several 'back of the envelope' kinds of sessions with him, too. Well, after *this* one, Bob said he thought there's a place again to do a really ballsy car. We had a number of things going, in the backs of our minds and on the backs of envelopes, so this was a perfect chance to come out of the closet. A couple weeks later, we had some full-size drawings and a bunch of sketches. I was just kind of testing how serious Bob was. As it turns out, he was very serious.

"We started up a full-size clay model. The Cobra as a role model was definitely an influence. What we're trying to do with Viper, though, is recreate Cobra in the way you *remember* Cobra, not necessarily its reality. The front end is definitely Dodge-signature, yet we wanted to have a very oval look, something that was a descendent of race cars. We were specific in searching for a look on the front that was kind of sinister. Even the round lamp, the driving lamp, was there right from the beginning.

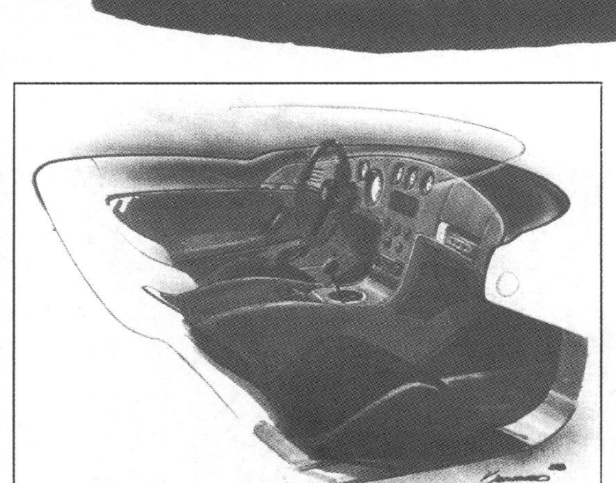

"I wanted side exhausts because you look back at the Corvettes of the mid-'60s, and they have it. The biggest challenge was the catalyst. The catalysts reside in the sill.

"Literally everything we could find [to start up development], we stole. It's not often that projects have this kind of emotion. One reason we're able to translate the final appearance so closely to [the concept] is that there is room for the bumper beam with really subtle modifications to this grille opening. And of course, the rear offset is fairly close. The concept wasn't cheated in that regard. The place where we probably took the most license was in the windshield, but the angle, the cowl touchdown and everything, was adopted. The surface of the original concept car and the final car are the same."

CHRYSLER'S ALUMINUM V-10

THE HEART OF THE VIPER

by Don Sherman

Shortly after the Dodge Viper design study appeared on the auto-show circuit in January 1989, more than a few premature deposit checks were mailed to Chrysler's Highland Park, Michigan, head office. As it turns out, customers were *not* stirred to a check-writing frenzy solely by the Viper's sensuous open-roadster curves. Raw sports-car sex appeal has been readily available in America for years. Rather, the Viper's unique feature and the prime attraction to those eager for a place on the production-car waiting list is a thumping 8-liter V-10 under the muscular hood.

The idea of powering the Viper with such an extraordinary engine had an appropriately unusual origin. Several years ago, Dodge Ram pickup trucks fell into a sales slump. The Chrysler team dispatched to analyze the problem discovered that the Ram's most powerful 5.9-liter V-8 engine wasn't potent enough to fulfill certain customer needs. Chevy and Ford pickups both offered big-block V-8s with more than seven liters of piston displacement to tow extra-heavy loads, such as horse haulers and travel trailers.

Unfortunately, there was no easy way for Chrysler to add muscle to its Ram pickup truck. The tooling that once produced the corporation's big-block V-8s had long ago been scrapped and the 5.9-liter small block could not be stretched far enough by conventional means to match the torque curves available in Chevy and Ford pickups.

To save the day, Chrysler President Bob Lutz concocted a bold but wacky solution: Why not simply tack an extra pair of cylinders onto the 5.9-liter V-8 to create a strapping V-10 that could be manufactured on existing tooling? The naysayers promptly stormed out of the woodwork. Technical foes insisted that a 90° V-10 engine would be horribly out of balance; shaking forces would be unbearable, even in a truck, they claimed. Marketing opponents declared the idea too radically different from the conventional Chevrolet and Ford approach and that the truck-buying public would never accept such an oddball engine design.

Lutz, however, persisted. He pointed out that 5-cylinder engines by Audi and Mercedes-Benz were perfectly acceptable, that the V-10 was an established truck engine configuration in both Europe and Japan. Eventually, he gained one ally, then more. An experimental V-10 engine was created by literally brazing together portions of the existing V-8's castings. When it was first fired up on the dynamometer, it ran more smoothly than expected. There were no serious vibration problems. Lutz and his supporters were vindicated.

There are several reasons why no auto manufacturer has recently adopted a V-10 layout. Such engines are inevitably heavy and bulky. For optimum vibration characteristics, the proper angle between cylinder banks is 72°, another strike against Chrysler's approach. In spite of these drawbacks, this layout may be gaining in popularity. Honda, Renault, Judd, and Alfa Romeo have all designed V-10s for Formula One racing, and the all-new '91 Acura Legend Coupe may be powered by such an engine when it debuts this fall.

Chrysler's V-10 configuration for the Viper and future Dodge truck engines is not ideal, but there appears nothing to keep it from doing a remarkable job pulling horses and producing horsepower. Essentially, this is an all-new design that retains the 5.9-liter V-8's bore centers and deck heights. Since the cylinder-bank angle is 90° instead of the theoretically correct 72°, there's an uneven firing pattern. The interval between power pulses alternates between 54 and 90° of crankshaft rotation. Primary forces are in balance, but a secondary lateral shake exists, which, according to a Chrysler engineer, is smaller in magnitude than the typical V-6 disturbance.

The Chrysler V-10's cylinder block has a number of heavy-duty features in keeping with its intended application: free-standing cast iron cylinder liners, ladder-type main-bearing girdle, and deep side skirts with a wavy surface to minimize radiated noise. The cylinder heads have an advanced high-efficiency combustion chamber and a conventional pushrod valvetrain in keeping with the Viper's "retrotech" image. Nevertheless, a great deal of contemporary technology is incorporated in the V-10's design: distributorless ignition, roller-type hydraulic valve lifters, tuned intake and exhaust plumbing, sequential port fuel injection, computerized engine management, single-belt accessory drive, and aluminum castings for all major components. (While Lamborghini assisted in the conversion of the design to aluminum, production castings will be sourced in the U.S. Truck versions of the engine—not due for several years—will most likely use a cast iron block and heads.)

With a 4.00-in. bore and a 3.88-in. stroke, the Viper's virile heart displaces a full 8 liters (488 cu in.). Merle Liskey, executive engineer in Chrysler's Jeep Truck Engineering department, asserts that there will be no problem meeting the Viper's conservative output projections: 400 hp and 450 lb-ft of torque.

Vibration created by the uneven firing order is clearly not a problem. Liskey claims the aluminum V-10 weighs about the same as an iron V-8. The exhaust note sounds more serpentine than spirited, but the Viper hasn't yet undergone voice training. That procedure may not be necessary: When the accelerator is mashed, the howl of protest by the rear pair of Goodyear Eagle GTs will doubtlessly drown out all lesser sounds.

The Dodge Viper has a six-speed gearbox, an 8-litre engine and 10 cylinders. It will reach 100mph in 10secs, weighs only a fraction more than a Ferrari 348 and it looks like nothing on earth . . . unless you remember the AC Cobra. Phil Berg takes the wheel of the world's most spectacular new sports car

ADULTS ONLY

FRANCOIS CASTAING, CHRYSLER'S chief engineer, describes the Dodge Viper as, "not a car for little children".

With that warning, we advise readers not to let recent offspring read our immediate thoughts about the car pictured here: the Viper is the most hellacious, visceral roadster we've ever experienced. And it's the experienced, not the immature, which this wild 188mph topless muscle-car is intended to please. It has only two seats anyway, so leave the children at home.

The Stetson-wearing sheriff who stopped Chrysler president Bob Lutz for driving at 120mph on a fast Arizona country road didn't think the Viper was a plaything either. The lawman was hurrying his police Thunderbird at 85mph on his way to another incident when Lutz passed the unmarked Ford like it was a tourist's caravan.

Passing the sheriff was easy: the Viper is a 400bhp, 8-litre, V10 roadster that Chrysler will build and sell in the US by 1992. Sprinting to 100mph takes just 10 secs.

"The sheriff said to me: 'You can't drive like that. I was on call and *I* don't drive like that,'" Lutz relates about the confrontation with the lawman. "He didn't give me a ticket because it was just so unbelievable."

Front-engined, rear-driven and with a glass-fibre body on a steel-tube frame, the Viper looks like a car of the future, with its low-profile, huge tyres, and squinting high-tech headlights. And it also looks like a blast from the past, with its distinct Cobra-like layout. But it's all new, designed from the beginning to have prodigious power.

The Viper's spectacular looks came in 1987 when Chrysler's director of design, Tom Gale, came up with a rendering of what a modernised version of the famous 1965-1967 AC/Shelby Cobra would look like during a meeting with chairman Lee Iacocca, Castaing, Lutz, and racer/car builder Carroll Shelby. The executives liked what they saw and the first show car made its debut in Detroit in 1989 to overwhelming enthusiasm.

Three other complete cars have now been built to refine suspension tuning and the V10, which has been under development for the company's new pick-up truck not due until 1994. The Viper show car sported exposed exhaust side pipes immediately aft of the front wheels, but the running prototypes have only a vent in the front quarter panel, and ▶

Not a car for children, warns chief engineer Castaing (left). With 400bhp and 188mph top speed, Viper is strictly for experienced drivers and enthusiasts

Chrysler's Bob Lutz (standing) discusses Viper details with engineer Dick Winkles. Looks came from design director Gale's concept of a modernised AC/ Shelby Cobra. Interior is spartan with no room for kids

◀ side pipes under the doors. Road-going Vipers also have targa bars behind the seats, an addition made partly for aerodynamics and partly for body strength. Some states have regulations against side exhaust pipes, so Chrysler is also developing a version with dual rear exhausts.

The biggest problem is noise. At speed, from the driver's seat, you hear only the exhaust pulses from five of the car's 10 cylinders. The noise rumbles up from an area level with your elbow. Some Chrysler executives have said the Viper sounds like a truck with the side pipes. Considering the state of high-powered pick-up trucks in the US, like Dodge's 5.2-litre V8 Dakota and Chevy's roaring 7.4-litre 454SS, the correlation is not insulting. Both Viper versions are a massive undertaking by Chrysler engineers, who are working against time to get production-ready cars by 1992.

The car pictured here is prototype number two. It is fast, loud, and more refined in chassis and aerodynamics than its newness would suggest. This is not a cobbled-together kit car, nor a race car with a body, but a real road car capable of tremendous performance with as much comfort as a Corvette convertible.

Climbing inside is easy enough. The doors look small, but the car sits high so that you feel you're entering a standard saloon. There is plenty of legroom in front of the driver's Recaro seat, but tall passengers must sit with knees bent. A four-point harness holds you tight in place and the upper door curves around almost to your shoulder.

The 7990cc V10 starts instantly. It is controlled by two electronic brain boxes from Dodge's V6 Dakota, but Viper engineers explain that a single box is being developed. Idling, the huge engine makes low, guttural noises, not pulsing like a V8 hot rod motor, just a continuous baritone hum. Much of the power

DODGE VIPER V10

LAYOUT
Longitudinal front engine/rear-wheel drive.

ENGINE
Capacity 7990cc, 10 cylinders in vee.
Bore 101.6mm, **stroke** 98.6mm.
Compression ratio 9.5 to 1.
Head/block al alloy/al alloy.
Valvegear ohv, 2 valves per cylinder.
Fuel and ignition Electronic ignition and sequential fuel injection.
Max power 400bhp (PS-Din) (298kW ISO).
Max torque 450lb ft (610 Nm).

GEARBOX
Six-speed manual.
Final drive ratio 3.07, limited slip differential.

SUSPENSION
Front, independent, double wishbones, coil springs, telescopic dampers, anti-roll bar.
Rear, independent, double wishbones, lateral links, coil over dampers, anti-roll bar.

STEERING
Rack and pinion.

BRAKES
Front, ventilated discs.
Rear ventilated discs.

WHEELS AND TYRES
Aluminium alloy, front 11ins, rear 13ins rims, Goodyear Eagle ZR, front 275/40ZR17, rear 335/35ZR17 tyres.

DIMENSIONS
Length 175.1ins (4447mm) **Width** 75.7ins (1923mm) **Height** 44.0ins (1118mm)
Weight 3300lb (1498kg).

PERFORMANCE (claimed)
Manual 0-60mph 4.0secs. Maximum speed 188mph.

A wild car in wild country: Lutz loves to take the Viper for a spin in Arizona. Steel-tube frame is rigid, providing firm base for glass-fibre body. Car is front-engined rear-driven

is delivered below 5000rpm and throttle response is instant.

Underway, the sound of the V10 completely overwhelms any wind noise. You're aware of wind coming over and around the sides of the short-screen from 60mph. At more serious speeds, you lean towards the steering wheel and the wind noise disappears.

The Viper has amazing grip from its 275/40Z-17 front and 335/35Z-17 rear tyres. The car is stable, and body roll is nearly non-existent although it has a soft ride for a big roadster, much like a Jaguar XJ-S convertible. The chassis is very rigid and, because of that, the glass-fibre body doesn't squeak or crack on bumps and potholes.

There was some concern over the Viper's aerodynamics during its development. The first prototype resembled more an aeroplane's wing than a motor car and, according to Lutz, had about 450lb of lift at 170mph. But this prototype, he says, "has *no* lift". Straight-line aerodynamics are very good on this car, he claims, though he remains concerned about crosswind stability at very high speeds.

Chrysler has built a replica of some sections of Germany's Hockenheim grand prix circuit at its Chelsea proving ground in Michigan, and Lutz says he's driven Vipers at 140mph on the course to check stability. Formerly Ford's top man in Europe, Lutz says the German racetrack is one of his favourites. "In European-style hillclimbs, this car will wax anything. You can bring the tail out with power . . . it's magic." But Lutz also likes driving the car in Arizona, despite the attention from the local law-keepers. "The car feels so much better on the road than at Chelsea."

The front-engine layout was favoured not only because it was true to the original Cobra, but because Lutz happens to be six feet, three inches tall. "I don't fit in a mid-engined car," he says. Lutz owns an AC Mk IV and compares driving that against this Viper: "My Mk IV bottoms often on bumpy roads," he says. This prototype Viper, with 1100 miles on its odometer, doesn't bottom its suspension at all. In fact driving at 110mph down a steep hill with a cattle-crossing grate at the bottom, the Viper's frame scraped the asphalt, before the suspension could bottom. "The suspension is pretty final," he adds about the red prototype.

The target weight for the car had been 3300lb, about the same as Ferrari's 348 tb, but this Viper tips the scales at 3450lb, the same as the ZR-1 Corvette. The engine in this car is made from iron, but there is an aluminium engine ready for the next prototype which will shave off some weight. It will also provide a 50/50 front/rear weight distribution, the current car being fractionally front-heavy. The aluminium engine develops 496lb ft of torque, of which the bulk is produced between 2400 and 3600rpm. The iron block is still good for 450lb ft, according to Dick Winkles, the bright, young Chrysler engineer in charge of developing the Viper engine.

A further 20 prototype Vipers are planned, first for endurance testing, then for the US Environmental Protection Agency certification tests for fuel consumption and exhaust emissions. After that, Department of Transportation crash testing will destroy a few more cars. A handful of pilot production cars will come next. The team responsible for designing and engineering these Vipers is made up of just 50 people, dedicated to this task.

Chrysler has been looking at places to build the Viper and suggests that it will be built in Detroit, at a stamping plant the company operates in the city. Although Chrysler would have preferred to use a greater number of previously developed parts on the car, the only pieces it took off the parts shelf were a steering column from the Jeep Cherokee and the front upper control arms from the Dakota pick-up. Everything else is handmade and will have to be produced new.

Chrysler expects to sell up to 5000 Vipers a year, but can build more, according to Lutz. Lutz is also careful to mention that he doesn't want the roadster to sell in large numbers, in order to keep it exclusive. Since Chrysler has never built a sports car before, it will take some time to become profitable. GM executives have said the Corvette, which began production in 1953, took about six years before it began making money. A small volume for the Viper, then, would lessen the financial risk. Chrysler, aware that its image in the US is floundering and that its business is largely minivans and Jeeps, says it is committed to the Viper.

We hope so. The level of development already in the car proves it's not just a publicity stunt. This is a real car. A real fast car.■

COBRA VS VIPER

SNAKES, RATTLE & ROLL

BY RAY THURSBY

PHOTOS BY LESLIE L. BIRD & DEAN SIRACUSA

Words fail me. Descriptive phrases, neatly turned sentences that make dull cars exciting and exciting cars irresistible, just don't fall readily to mind when it's time to make some sense of the meeting between a 1966 Shelby Cobra and a 1992 Dodge Viper. Call it sensory overload, numbness, that period of tingling nerve ends and jumbled thoughts that inevitably follows a bungee jump, a trip in an F-15 fighter plane or a ride in what may be the most exciting performance machines that ever pounded pavement.

Yes, I know there are cars that can outrun the Super Snakes. I've ridden in—and even driven—a few myself. But that's beside the point. Nails can be driven with the smallest of hammers; a 10-lb sledge makes a bigger impression, even though it's doing the same thing.

Perhaps it is the tightly focused single-purpose nature of these two beasts. Each has an oversize engine, enough pieces to get the engine's power to the ground, a rudimentary body, two seats and a steering wheel for the driver to hang on to. No elaborate radios, no electric windows—hell, no windows at all, unless you want to attach them yourself—and only the most minimal concessions to civility as demanded by bureaucracy.

Despite the differences in number of cylinders under the hood, appearance and specification, these are the same car at heart. Neither bears any resemblance at all to the assembly line products of manufacturers' bread-and-butter teams; they carry the stamp of the individuals who created them.

To a large degree, both cars owe their existence to Carroll Shelby, one of the least timid of men. Don't let the carefully created and well-honed image of a good ol' Texas boy wearing bib overalls deceive you: Shelby is one shrewd

■ This 427 Cobra sports a rare intake manifold, which accommodates four Weber 52-mm IDA carburetors. Halibrand mags help put more than 400 bhp down to the ground.

cookie, not averse to earning money or getting publicity for himself and/or his products.

Which financial rewards and publicity, let it be said, have been earned. Shelby is, after all, the one who thought up the Cobra, and got it into production. Before the Cobra era, he was a talented and successful racing driver, afterwards a builder of high-performance Chrysler products, with a few side ventures thrown in for good measure. It wouldn't do to underestimate the man or his accomplishments.

The Cobra story has been told any number of times, has been the subject of exhaustive scrutiny; there isn't much to add at this late date. Shelby saw potential in the marriage of British chassis and American engine, and convinced AC and Ford to play along. Early Cobras, powered by the then-new Ford 289-cu-in. V-8, proved his initial assessment right.

But the early Cobras had their limitations, centered mostly around their transverse-leaf spring suspensions. No real surprise that a chassis setup developed in the early Fifties for use with low-power engines was not quite up to even the 271 bhp of the basic Cobra engine.

Stuffing a 427-cu-in. V-8 into one of those early Cobras, as the late Ken Miles did under Shelby's direction, made matters worse. More speed was available, but making use of it was, to say the least, a tricky proposition. The final production 427 Cobra ended up with a new frame and a completely redesigned suspension. As such, it became a car that mere mortals could drive. If they were careful.

The Viper's engine has an additional two pistons and 500 cc over the Cobra's 427-cu-in. mill. Fuel injection allows more power and equal civility.

In the end, the Cobra was a success. It won races, made money for Shelby, got lots of favorable publicity for Ford, and staved off the collapse of AC Cars (which didn't occur until just this past year). Everybody got something out of the deal, not least among them the collectors who ended up with the original cars and the entrepreneurs who built a thriving business selling reproductions.

■

By contrast, the story of the Viper's conception and gestation has yet to be told. It has been said that the idea originated within the Shelby operation and that the concept, if not the first prototype, was transferred to a development group within Chrysler. Early rumors did indeed concern a "new Shelby sports car," and Ol' Shel himself was indeed on hand when the Viper was publicly unveiled. How it *really* happened is a tale that will have to wait for another time.

What is clear, however, is that the Viper was influenced by the Cobra. Its very concept, not to mention its name, represents unabashed homage paid to the ultimate Original Snake of 25 years ago.

Not a bad base on which to build.

Shed of their skins, both snakes can be described thus: A multi-tube structure forms the basic chassis, carrying fully independent suspension. Within that frame, a large iron-block pushrod engine is inserted, driving the rear wheels. Disc brakes are fitted at each corner, and curb weight has been kept below 3000 lb. From there, the two cars' makeups diverge.

Before we go any further, though, please understand that the Viper is a prototype and that many details of its appearance and specification are still being reviewed. It seems certain that Chrysler wouldn't dare put it on the market with a turbo-4 under the hood, but individual items are still subject to change.

Ford developed the 427 (7.0-liter) engine used in the Cobra for performance. Installed in Galaxie sedans of the day, it became a fixture in NASCAR racing, and powered the later Ford GTs that so effectively trounced Ferrari in sports car competition. It was a passenger-car engine massaged to provide the ultimate in horsepower at a time when GM and Chrysler were mounting race track challenges with their own race-worthy V-8s. As provided to Shelby, the 427 developed 425 bhp and churned out 480 lb-ft of torque.

The Viper's engine is not likely to end up under the hood of a family sedan. It is a truck

engine, destined for use in upcoming Dodge utility vehicles. Though 8.0 liters is nothing to sneeze at, what sets this engine apart from other production powerplants is not so much its displacement as its configuration: 10 cylinders, arranged in vee-formation. Henry Ford messed around with prototype V-10s more than 60 years ago, but Dodge's will be the first one available to the public. In current form, the Viper V-10 puts out an estimated 400 bhp and 450 lb-ft of torque. With Lamborghini's help, Dodge is preparing an alloy-block version of the V-10 for production Vipers. Nevertheless, the cast-iron device used in this prototype, while perhaps too heavy and prosaic for the customers, seems appropriate for this story.

Manual transmissions are found in both cars. Cobras were fitted with Ford-built 4-speed gearboxes; the Viper unit has two additional ratios and comes from Getrag. Floor-mounted shifters are the rule here, with the edge going to the Viper's straight lever over the Cobra's curiously bent and somewhat awkward fitting one.

Clothed respectively in aluminum and fiberglass, the Cobra and Viper can easily be told apart with a single glance. The Cobra's body shape was old when the car was new, having been developed from a design that began as a copy of the Touring-body Ferrari 166 Barchetta. As engine and tire sizes grew, it was stretched in every direction to meet new requirements. Small chrome-plated "nerf bars" provided all the bumper protection then thought necessary. It remains a handsome, muscular, classic shape.

The Viper owes its body design largely to some overheated imaginations at Chrysler. From front to rear it sweeps, swoops and swirls around in a manner far too assertive for application to everyday cars. Being lower, wider and longer than the Cobra, the Viper gave its stylists every opportunity to create dramatic visual statements with a minimum of hindrance from reality, which they did. I'm undecided about my impression of it; I *think* I like it, and the Viper will never be lost in a crowd.

■

Interior design of the two machines reflects a sameness of purpose and difference of birthplaces. The Cobra is a British sports car inside, with lots of instruments spread across a flat panel, barely acceptable seating for two, and a wood-rim steering wheel for the driver. Restyle the whole thing, add better seats and replace wood with leather wrapping, and you have the Viper cockpit. Taller occupants won't

SHELBY COBRA 427 PROVIDED BY STEVE BERRES, PHOENIX, ARIZONA

have to go through contortions to fit in the Viper as they must in the Cobra; but they will note that function has not been compromised. Advantage: Dodge (to coin a phrase), with honorable mention to the AC.

Barring the unforeseen, Dodge expects to produce an initial run of less than 500 Vipers for the 1992 model year. Quantities half and double that size have been publicly discussed, because no one really knows the dimensions of the Viper market for sure, but 500 seems a nice round figure. Price? The target number is around $60,000 which, given the enormous investment Chrysler must have in the project, is entirely fair. Granted, the Viper is meant more as an image-enhancer than a profit generator, but the stockholders will no doubt be happier if the project comes close to breaking even. Expect a long line at the Viper order desk.

Déjà vu time: Shelby built fewer than 400 Cobra 427s at a list price of $6995 each. At that, the last few were, some say, a hard sell. How times (and prices and attitudes) have changed!

On to the showdown: Snake vs Snake. If the Viper has a rival, the Cobra is it. None of today's high-tech marvels from overseas can do more than put up numbers against these Leviathans, and numbers mean little. This is a story of elemental performance, sheer ground-shaking power, having nothing at all to do with creature comforts or fuel mileage tests. The faint of heart and weak of limb would be most uncomfortable on this ride.

Ride is the operative word where the Viper is concerned. Civilians have not driven the Viper prototypes to date, but a few rides have been given out to journalists. Mine came with Chrysler honcho Bob Lutz in the left seat. Lutz is no stranger to high-performance machinery, and has had enough time in the Viper to feel at home with it. The evidence gathered this day suggests that drivers of more limited talent should approach the Viper with care.

Key turned, the big V-10 lights off with a rumble that is initially disconcerting to bent-8 fans. Logically, it sounds much like two Audi inline-5s coupled together. But once under way, its bark is remarkable.

So's the bite. We're quickly up to high speed,

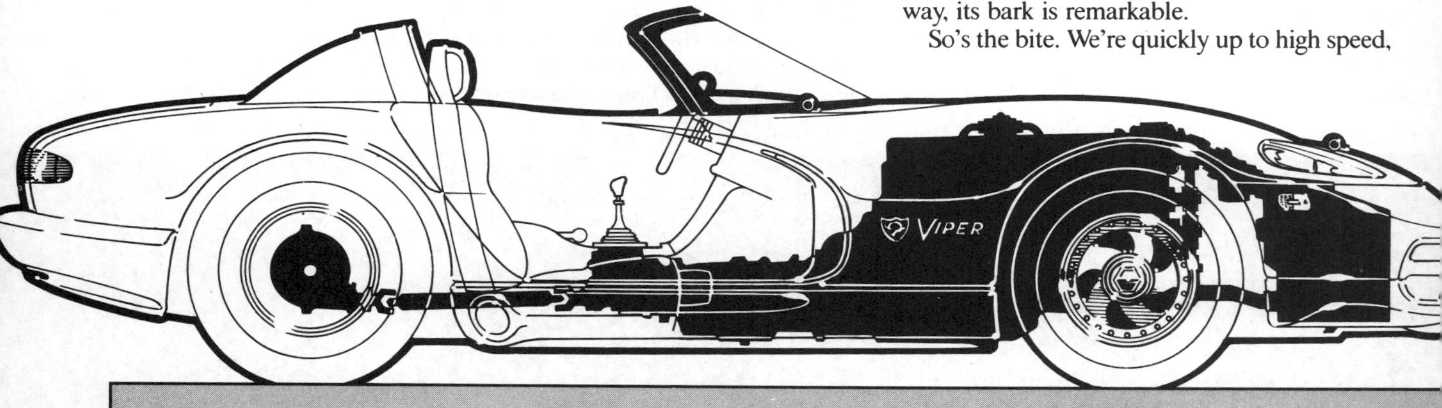

Dodge Viper

SPECIFICATIONS	
Price	na
Curb weight	2990 lb
Wheelbase	96.2 in.
Track, f/r	59.6 in./60.8 in.
Length	172.0 in.
Width	75.6 in.
Height	46.2 in.
Fuel capacity	na

CHASSIS & BODY	
Layout	front engine/rear drive
Brake system, f/r	discs/discs
Wheels, f/r	alloy, 17-in. diameter
Tires	275/40ZR-17 f, 335/35ZR-17 r
Steering type	rack & pinion
Suspension, f/r:	upper & lower A-arms, coil springs, tube shocks, anti-roll bar/multi-link, coil springs, tube shocks, anti-roll bar

ENGINE & DRIVETRAIN	
Engine	ohv V-10
Bore x stroke	101.6 x 98.6 mm
Displacement	7989 cc
Compression ratio	na
Horsepower, SAE net	est 400 bhp @ 5500 rpm
Torque	est 450 lb-ft @ 2200 rpm
Fuel injection	electronic port
Transmission	6-sp manual
Final-drive ratio	na

PERFORMANCE	
0–60 mph	est 4.1 sec
Standing ¼ mile	est 12.5 sec
Top speed	est 190 mph

na means information is not available.

fast enough to permit use of the Getrag 6-speeder's long top gear. Ride and handling qualities are, as you'd expect, firm and precise, biased toward maximum grip and stability.

It is comfortable in the Viper (and drafty, of course); it looks as wide from inside the cockpit as it does from the outside. The prototype's 5-point seatbelts are much appreciated at 3-digit speeds as is the effectiveness of the four giant disc brakes. Impressive. (As an aside, Lutz confesses to wishing the Feds would allow 5-point seatbelts; it seems, alas, that passive restraint belts are inevitable. Phooey.)

The Cobra is louder than the Viper, just as drafty, feels faster, rides even more harshly and is equally wonderful. The noise from under the hood drowns out, well, everything. There seems to be more feedback through the steering, possibly less ultimate grip in corners, which is logical when you consider the years of technology that have come between Snakes.

Though my Viper experience was confined to long straights and gentle, sweeping bends, I sense in the car the potential for becoming a double handful on twisty roads. I know that is true of the Cobra. There's only the finest of lines—and the slightest change in right foot pressure—between being on the ragged edge and being in way over your head.

Fortunately, this is one of those times when drawing a conclusion is unnecessary. I love the Viper and Cobra, and would cheerfully make room in the garage for either (or both). Under duress, I'd opt for the Original Snake out of pure sentiment, and because of a sneaking hunch that the Cobra, unfettered by the rules and regulations that affect the Viper, might be just a little hairier and faster.

But the Viper is what's coming up next; give the Iacocca and Lutz duo credit for that. There are no more new Cobras, just new Vipers on the way in less than a year.

Give most of the credit to Carroll Shelby, though. Seeing the two machines together, you'd think he never stopped building Cobras, but decided simply to name the latest version after a different snake. ■

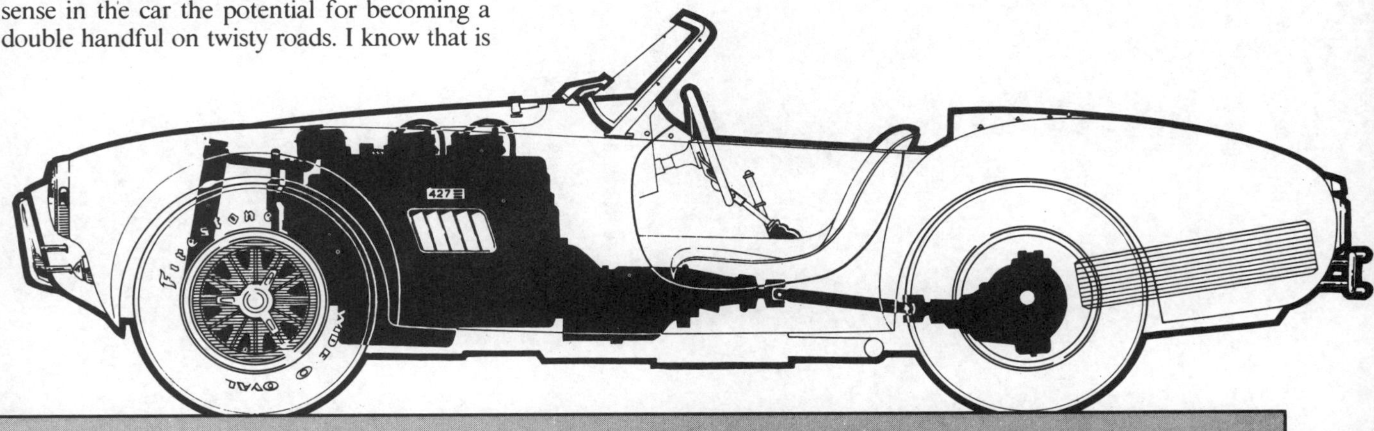

Shelby Cobra

SPECIFICATIONS	
1966 price	$7495
Curb weight	2530 lb
Wheelbase	90.0 in.
Track, f/r	56.0 in./56.0 in.
Length	156.0 in.
Width	68.0 in.
Height	49.0 in.
Fuel capacity	18.0 U.S. gal.

CHASSIS & BODY	
Layout	front engine/rear drive
Brake system, f/r	11.63-in. discs/10.75-in. discs
Wheels, f/r	cast alloy, 15 x 7½-in.
Tires	Goodyear Blue Dot, 8.15 x 15
Steering type	rack & pinion
Suspension, f/r:	upper & lower A-arms, coil springs, tube shocks/upper & lower A-arms, coil springs, tube shocks

ENGINE & DRIVETRAIN	
Engine	ohv V-8
Bore x stroke	108.0 x 96.2 mm
Displacement	6998 cc
Compression ratio	11.5:1
Horsepower, SAE gross	425 bhp @ 6000 rpm
Torque	480 lb-ft @ 3700 rpm
Carburetion	2 Holley 4-bbl
Transmission	4-sp manual
Final-drive ratio	3.54:1

PERFORMANCE[1]	
0–60 mph	5.3 sec
Standing ¼ mile	13.8 sec @ 106 mph
Top speed	162 mph

[1]from Cobra 427 road test in *Road & Track*, July 1974.

DODGE

FLAT OUT AT INDY

COVER STORY

VIPER

INSIDE THE CORVETTE KILLER

BY C. VAN TUNE

PHOTOGRAPHY BY ROBERT MARTIN, JIM FRENAK, AND THE AUTHOR

The wall on the outside of Turn One rushes up quickly at 150 mph, and we're doing every bit of that speed as we blast headlong down the front straight of the most famous racetrack in the world. Beside us, in the business seat of the muscular Dodge Viper, is Chrysler performance consultant Carroll Shelby. There aren't many 68-year-olds we'd feel comfortable with at these speeds, and certainly fewer yet who've had a heart transplant. But that's exactly the case with Shelby. One of the last living legends of our automotive lifetime, Shelby received the heart of a 35-year-old in June 1990. A mere 11 months later, he was driving the Dodge Viper Pace Car at the 75th running of the Indianapolis 500.

The Dodge Viper is every bit as miraculous an entity as is Shelby. Cars like this just aren't supposed to be able to exist in 1992 (when the first Viper will go on sale). But here we are, screaming around the Brickyard in a prototype of the sort of no-holds-barred, get-with-it-and-go-fast roadster the likes of which we haven't seen since the demise of the Cobra 427. That was in 1967. Today, some 25 years later, besieged by fun-stealing acronyms such as EPA, DOT, NHTSA, and CAFE, cars are supposed to have small-displacement inline four or V-6 motors, for crying out loud, not this gargantuan 8.0-liter (that's 488 cubic inches) V-10.

Ten cylinders. Dwell on that for a moment. More cylinders and ground-pounding cubic inches than any factory hot-rod from the much-heralded muscle-car era. And this low-compression, catalytic converter-equipped motor produces power levels just as impressive as any high-compression, high-octane, leaded fuel-sucking 426 Hemi or big-block Chevy from 25 years ago. Even with today's more conservative "net"

power ratings, it's good for 400 horsepower plus and 450 foot-pounds plus of torque, which equates to something like 475 horsepower and 525 foot-pounds when calculated in the old "gross" power ratings. Connect that level of thrust to a 3250-pound object, and prepare for acceleration that'd make an F-14 fighter jet seem weak by comparison. (In fact, engine development work was accomplished in Chrysler Engineering's "Dyno cell No. 13," the same dyno used to develop the 426 Hemi and NASCAR engines in the '60s and the only dyno in the company with the torque capacity to handle the V-10.)

By now, everyone's heard the story of how the Viper came to be. Less than three years ago, Shelby, Chrysler President Bob Lutz, VP of vehicle engineering François Castaing, and styling chief Tom Gale were in a meeting to discuss new performance projects. During the course of their talk, the subject drifted to great cars of the past, and, of course, the Cobra was high on that list. The bull session quickly turned serious when someone suggested Chrysler could build a modern version of the powerful two-seat roadster. Eyes widened, light bulbs clicked on, and pencils attacked paper. Gale's sketches grew into the Viper show car that debuted at the Detroit auto show in 1989 as a "styling study," built solely as a teaser, a measurement of public reaction to such a radical design.

THE LUSTY FEEL OF THE VIPER IS INTOXICATING AT HIGH SPEEDS, BEGGING YOU TO GO FASTER

Within days, Chrysler Chairman Lee Iacocca's in-basket was overflowing with letters from hopeful Viper buyers—many with $1000 deposit checks attached. Chrysler returned the checks, but knew it had too strong a prospect in the Viper to merely let it fade away with the innumerable show cars of the past. A team was assembled to study the feasibility of putting it into production. Development cars were built, and testing began. In May 1990, Iacocca announced that Viper would become a production reality and that it would debut at the '92 auto show—a mere three years after the first "styling study" was shown. Can you name another show car that made the transition from turntable art to production in a form as true to the original as the Viper is? You may see a headlamp treatment or greenhouse design carried over to production, but there hasn't been a car we can remember that bears such a close resemblance to the original design.

As amazing a story as the lightning-quick development schedule is the fact that a mere $50 million is the total development budget required to bring Viper to market. In these days when $150 or $200 million is considered a bargain program, the Viper's figure is tantamount to what you'd use for lunch money. The team approach (see "The Green-Flag Gang" sidebar) is responsible for much of this cost- and time-savings ability, and is a formula that's rapidly spreading throughout Chrysler.

The Viper is an excellent gauge with which to determine who are the pretenders among the worshippers. Although the car's head-turning abilities are second to none, we quickly heard from the wimps in the crowd when they learned of the Viper's level of standard equipment. "What, no air conditioning?" whined one. "Where are the power windows and cruise control?" bleated another. For the big shocker, tell 'em there are also no side windows, top, or lockable doors. Then watch about 90 percent of 'em leave, shaking their heads in disbelief. The Viper has no direct competition and doesn't try to emulate Corvette, Ferrari, or Porsche. It's the lone vehicle in a new automotive micro-niche that could be called "retro muscle."

Shelby and the members of Team Viper went to great lengths to protect the back-to-basics approach. Like the Mazda Miata, the Viper is a throwback to an era when sheer driving fun meant more than electric sun visors or power-operated ashtrays. Everything about the Viper is purpose built. Like a precision piece of athletic gear, the car is all function, no fluff. "We built a car like the Cobra, but updated it with '90s technology," Shelby explained. "We purposely stayed away from four-valve heads, turbos, or other complicated stuff. We wanted this car to be easy for the average guy to work on. And it is."

Granted, there's nothing high tech about the motor. The massive engine (aluminum block and heads) still uses pushrods for valve actuation, and the single in-block camshaft is profiled for boatloads of low-rpm torque. And like those muscle-cars we dreamed about in our youth, the Viper responds with instantaneous force. Merely breathe on the throttle, and you'll send the steamroller-size Michelins up in smoke. Sixty mph happens before you're out of first gear (in the low 4-second range) and the standing quarter mile, according to a Viper project supplier who drove the car, is a spine-compressing 12.4-second/114-mph ride. With no top or side windows, and being pelted by the raspy blat of the side exhausts, it's like electroshock therapy for the g-force impaired.

The lusty feel of the Viper is intoxicating at these speeds, begging you to go faster, faster, and Shelby responds with his famous grin that's as wide as a west Texas sunset. Ol' Shel's not babying this machine (as if he could do so even if that were the order) as we rocket around the 2.5-mile oval track that's actually more square than it is round and isn't as sharply banked as it looks on television. It also seems particularly narrow at our velocity of over 200 feet per second, as we run in the same groove the race cars have for thousands of laps.

THE VIPER IS AN EXCELLENT GAUGE WITH WHICH TO DETERMINE WHO ARE THE PRETENDERS AMONG THE WORSHIPPERS

Fortunately, we've got the security of five-point harnesses holding us in place. Production Vipers will use three-point door-mounted seatbelts, but will have mounting provisions for the harnesses. There's no airbag. Full analog instrumentation is used and consists of a 180-mph speedometer and 7000-rpm tach (redline at 6000 rpm) placed in front of the driver, with smaller canted dials for coolant temp, oil pressure, fuel level, and voltage residing left-to-right in the center of the dashboard. Beneath those instruments are a trio of air registers, the controls for the ventilation system, and (in the Indy Pace Car) the switches for the emergency strobe lights. Production Vipers will have a radio mounted in this location, which, with its six speakers, has been designed for "quality sound reproduction at 100 mph."

With the 10-cylinder engine's serenade, however, we can't imagine the radio getting much use. The speedometer and tach are hidden from the passenger's view, though, so you'll have to rely on the driver to keep you informed. "There's 120...now 130...now 140 mph," Shelby announces as we exit Turn Two and accelerate hard down the back stretch. The engine's torque curve is so flat the V-10 hardly seems to be working, even though we're accelerating at a rate only a couple production cars in the world could match. Wind noise and buffeting are surprisingly low, and the side exhausts are actually almost quiet at idle and in cruising situations. It's only when you open up the throttle plates and start to leg it that the temperament changes. The sound of the V-10 is unlike any other engine on the planet, with a full-out battle cry that'll freeze would-be challengers dead in their tracks.

Everything about the cockpit is basic and businesslike. The surrounds are far nicer than those of the barebones Cobra, however, as the Viper actually includes amenities such as door panels, center console, and carpeting. "The engineers put a lot of heat insulation in the footwells," Shelby explains, "so your tennis shoes won't stick to the pedals like they did in the Cobra." The highback leather bucket seats are supportive, comfortable, and large enough to accommodate those moderately wide in the beam. In fact, the interior is wider than you'd imagine for a car of

THE LINE FORMS HERE

With plans to build only about 250 Vipers during the first year of production, Chrysler will disappoint several thousand prospective buyers right off the bat. Although second-year production plans call for up to 5000 cars to be built, this atypical (for a domestic manufacturer) slow launch curve brings with it complications from day one.

Think of how you'd handle it if it were your company. You've got over 5000 people signed up on the "Viper Prospect" list, which is actually nothing more than a compendium of car nuts who'd sell their mothers for a Viper. Chrysler began recording names on this list last year, but being on it guarantees you nothing in the way of getting a car. Really.

Then, on the other side of the equation, you've got your 3000 Dodge dealers. Of which at least two thirds would sell their entire family to get one of these traffic builders in their showroom.

To be as fair as possible, Chrysler developed a system of checks to determine dealer eligibility. The three-test method looks at: (1) dealer CSI (Consumer Satisfaction Index) rating, (2) geographical location (major markets are far preferable), (3) sales volume (the bigger the better). From that equation, the division will determine which dealers get the initial 250 cars. But the dealers are completely autonomous in their decision to add any price premiums on the Vipers they receive. So, sad as it may be, there won't be many first-year Vipers selling for list. (Which is expected to be approximately $55,000, plus gas-guzzler and luxury taxes.)

By this point in time, it also seems useless to try to wedge your name onto the Viper prospect list. Ahead of you are politicians, movie stars, captains of industry, and other people with more money than most Latin American countries. And they honestly have no advantage over you in getting a Viper. *—C. Van Tune*

SHELBY AT THE WHEEL

There's been a lot of progress in the past 30 years, but some things you just shouldn't mess with. The Viper is the best of both the past and the future. It's all the wind-in-your-face fun the Cobra was, but updated with better brakes, tires, and other items that make it run better and go faster. When we started this project, some folks said we couldn't do it, do things like emissionize an 8-liter engine, or get the side exhausts to pass the government drive-by noise standards. But we did. And it's that sort of dedication that's made the Viper what it is today.

I put over 1000 miles on the car at Indy practicing for the race and giving rides to people. And it ran flawlessly That big V-10 has so much torque you can get nearly as much acceleration by short-shifting at 3500 rpm as you can running it up to redline. You can start out in most any gear, and there's good power even in fifth at as low as 1200 rpm. It's geared tall for CAFE (corporate average fuel economy) reasons, and 60 mph in sixth gear is only 1100 rpm, but I averaged better than 12 mpg during that thousand miles, and I sure wasn't takin' it easy.

We've got a really nicely balanced chassis in the car, with 50/50 weight distribution. It corners and steers well, and we've seen 1.0 g on the skidpad. The curb weight is 3200 pounds currently, but we've got some tricks up our sleeve to get it down to around 3100. We're using huge disc brakes, but without an anti-lock brake system. In our tests at Chrysler's Desert Proving Grounds, we convinced ourselves that anti-lock isn't necessary in a car this well balanced. I made two laps of the banked oval track at a constant 140 mph and holding the brakes on to build up heat. I then made a 120-0-mph hard stop and recorded absolutely no fade! The stop was straight and easy to modulate, as well. So, by eliminating ABS, we save weight, complexity, and cost.

In my opinion, the Viper is more just plain fun to drive than a $500,000 Ferrari. It's not supposed to be the world's most sophisticated sports car, just a '60s-style all-American performance car that'll put a grin on your face when you drive it. I'm truly proud to have worked with Bob Lutz and the rest of Team Viper on a car that people said couldn't be done. *—Carroll Shelby*

this size, and feels far more commodious than a Corvette convertible. You sit upright and tall in the Viper, in dramatic contrast to any of the Italian exotics, which adds to the Cobra-esque demeanor of this machine.

"I feel as if I'm in my coffin when I try to drive one of those Italian jobs," Shelby groans. "I hate to lay down and try to peek out of a little slit of a windshield. We designed the Viper to be a good old open-cockpit sports car like the Cobra was—a fun-to-drive car with unbeatable stoplight-to-stoplight performance."

In that fashion, too, the Viper is very much like the Cobra. "We didn't build this to be a 200-mph car," Shelby tells us. "I have no interest in driving that fast, even on the autobahn, and get a much bigger kick out of a machine that knocks your hat off at lower speeds. Zero to 100 mph and back to zero again. That's the kind of performance a guy can really use." (The Cobra 427 held the production car crown for its 14-second 0-100-0-mph talents. The Viper is reputed to be able to beat that performance.)

Don't get the wrong impression about this machine. It'll do 100 mph without breathing hard and (with its Borg-Warner T-56 six speed's 0.50 ratio top gear) has the capability to see the far side of 180 mph. But Shelby's point is well taken. Outside of a racetrack, where can anyone in this country hope to safely explore a 200-mph car's performance capabilities? A far better solution than owning a high-strung exotic with horrible city manners is exactly what the Viper delivers: instantaneous power from off-idle engine speeds and enough

torque to win any argument. People talk about horsepower, but what they really mean is torque—the force that shoves you in the back when you mash the throttle. There's no waiting for a turbo to spool up or for the engine to wind tight enough to "come up on the cam." Any time, at any engine speed, just step down on the loud pedal and hang on for mercy. You've heard the racer's maxim before: There's no substitute for cubic inches.

In the Cobra days, Shelby had a trick he'd pull on first-time riders. He'd tape a 10-dollar bill to the glovebox then tell the passenger to grab for it when he said "go." At that

STEP ON THE PEDAL, AND HANG ON FOR MERCY

THE GREEN-FLAG GANG

Three basic Viper precepts are:

- The engineering team will use a skunk-works approach and tap Chrysler resources and expertise while avoiding Chrysler bureaucracy.
- The engineering team will be volunteers with a strong vehicle focus and the responsibility to make quick decisions.
- The engineering team will act like a small company owned by team members who have mortgaged their homes.

In other words, Team Viper will, whenever possible, comport itself exactly like a small racing team operating independently within Chrysler Corporation.

Actually, that's quite possible because many Team Viper members are racers and have been for years.

Consider Pete Gladysz, the Viper chassis manager. While program manager for the original Shelby Charger in 1982, Gladysz "decided we needed to go racing to demonstrate the performance of our new product." Three of the first half dozen Shelby Chargers built were yanked from auto-show duty and hurriedly prepared for the '83 24 Hours of Nelson Ledges showroom-stock endurance race. One finished fifth overall.

In the '85 season, Gladysz and his small but growing gang of enthusiastic Chrysler engineers upped the ante with turbo power. The timing was perfect because Chrysler's turbocharged 2.2-liter four-cylinder was slated for widespread use in the future, and the powertrain engineering department needed a severe durability test. According to Gladysz, "Team Shelby's race cars became rolling test beds. We tried out engineering upgrades and turned in data gathered at the track. A 24-hour race is a worst-case test of any car's durability."

The effort was funded by powertrain engineering, marketing, and a lengthy list of vendors cajoled by Gladysz on the telephone. Chrysler engineers, technicians, and mechanics volunteered their evenings and weekends to prepare and campaign three cars. Several drivers were also Chrysler engineers. By season's end, the balance sheet was impressive: At a total cost of $350,000, Team Shelby entered three cars in six endurance events, finished every car it started, won three races, finished second three times, and raked in $20,000 in prize money. One Shelby Charger won the showroom stock A championship and team driver Neil Hannemann seized the drivers' laurels. A long list of engineering upgrades—ranging from added wheel studs to tougher transmission gear teeth—came directly from the racing experience. Powertrain engineering authorized large-scale production of the turbocharged 2.2-liter engine confident in its durability record.

Led by Gladysz, Team Shelby repeated this success in 1986 and then shifted its hard-earned front-drive expertise to IMSA GTU. Gladysz, who still dabbles in Pro Rally competition, is by no means unique, as shown by the following Team Viper roster:

Bob Lutz, father of the Viper, Chrysler Corporation president, and ranking member of the technical policy committee that oversees Viper development, was once a Team Opel rally driver and later piloted a few hairy BMW and Ford factory racers in "executive" events.

François Castaing, Chrysler's vice president of vehicle engineering and also a member of the technical policy committee, was in charge of Renault's successful Formula One turbo-engine program.

Carroll Shelby, LeMans winner and two-time Indy 500 pace car pilot, is Chrysler's performance consultant working with Team Viper.

Neil Hannemann, successful road-racing driver, is a Viper engineer concentrating on chassis, handling, and tire development.

Don Jankowski, a crew chief in the Team Shelby era, is a Viper development engineer overseeing manufacturing methods.

Dick Wrinkles, Team Shelby engine guru, is in charge of Viper engine performance.

Al Fields, former Team Shelby crew chief and a current road racer, is a Viper cooling-, fuel-, and exhaust-system engineer.

Ken Nowak, Team Shelby crew member, is responsible for assembly of Viper prototypes.

THE DEEP DYNO DOSSIER

The Viper V-10 began life as a spinoff of a future truck engine (which itself grew from Chrysler's 5.9-liter V-8), but development has taken this powerplant well beyond those humble beginnings. Lamborghini assisted the conversion of the block and cylinder heads from cast iron to aluminum and also revised the original port and combustion-chamber designs. A few magnesium castings have been added—valve covers and two accessory brackets at the front of the block. And modern technology in the form of roller lifters, distributorless ignition, sequential electronic fuel injection, and single-belt accessory drive will help meet Team Viper's ambitious weight and output targets.

Free-standing cast iron cylinder liners, forged pistons, and a forged-steel crankshaft are part of the program. With a 4.00-inch bore and a 3.88-inch stroke, piston displacement is a tire-torturing 8.0 liters (488 cubic inches). Casting, machining, and assembly processes are unique to the Viper V-10, and its connecting rods are the only significant components shared with the truck engine.

With a bare engine weight of 630 *pounds*, the Viper V-10 will never be taken lightly. A redline of 6000 rpm is anticipated, pushrods and all. Preliminary reports from the dynamometer room list 414 horsepower at 5200 rpm and 485 foot-pounds of torque at 4000 rpm (versus targets of 400 horsepower and 450 foot-pounds). Chevrolet, for one, hopes to join the Viper V-10 in the over-400 club sometime soon with a more potent version of its LT-5 Corvette ZR-1 V-8.

—Don Sherman

Brad Dotson, another crew member, works on Viper brakes.

Danny Houk (since deceased), was a tireless Team Shelby mechanic and also a Viper mechanic at Chrysler's proving grounds.

Charlie Brown, a member of the Viper engine group, currently campaigns an SCCA D-Sports racer.

John Donato, ardent drag racer, is a Viper powertrain engineer concentrating on transmission, rear axle, and driveshaft development.

Ray Schilling, a mechanic from Team Shelby days, is a Viper mechanic.

Bill Adams, another Viper mechanic, is also a drag racer.

Dave Buchesky, vintage racer, is a Viper engineer responsible for suspension and brake systems.

Jim Broske, SCCA road racer, is a Viper engine-development engineer.

Bob Zeimis, who dyno-tested Team Shelby's engines, now wrings out the Viper V-10 on the dynamometer.

Since Team Viper is but 60 members strong, it's clear that the racing mentality is well represented. Gladysz explains why this is essential to the cause: "Many of us have worked closely together for seven or eight years. We know what the next guy needs and how he operates. That's vital in assuring that a small, closely-knit group works at maximum efficiency.

"In the Team Shelby days, you could hand a guy a job and never worry about it getting done. That's the exact same attitude we have at Viper."

—Don Sherman

same instant, Shel' would nail the throttle, throwing the hapless rider back into the seat with so much force he could barely breathe, let alone reach forward and grab a piece of paper. "No one ever got that ten bucks away from me," Shelby chuckled. "With the Viper, I ought to use a hundred-dollar bill."

On the track, the Viper feels solid as a fortress, with absolutely no hint of cowl shake or chassis flex. The steel tube frame connects to tubular unequal length A-arms with coil-over shocks and melds to the pavement with 275/40ZR17 (front) and 335/35ZR17 (rear) radials. Behemoth four-wheel disc brakes are used, but anti-lock is not fitted (see "Shelby at the Wheel" sidebar for details). Power-assisted rack-and-pinion steering is about the only concession given to convenience items, as everything else is manually operated.

The body is constructed of a resin-transfer molding material that's a first for Chrysler, and all the first-year Vipers will be painted red. After that, other colors will be added, including black and possibly yellow, blue, and white. Like the Cobra, the Viper has a trunk capable of holding two golfbags (a design criteria). Also like the Cobra, an accessory soft top with slide-in side curtains will be optionally available, but don't count on it having the weathersealing ability of a Mercedes SL. And for the three or four gold-medallion types who really think they want a Viper, air conditioning will be a dealer-added accessory. Just so their silk shirts won't get too soggy on hot days.

Shelby pulls the Viper sharply off the line as we exit Turn Four and heads down into pit row. It's a narrow opening into the pits at 120 mph plus, and I can only imagine how difficult it must be to thread the needle at Indy race speeds. The braking force is practically choking the wind out of me as I strain against the shoulder harnesses, and I sneak a glance over to the heart transplant recipient to see how he's doing. "Relax," Shelby said with his familiar Texas drawl. "The Speedway doctor told me my heartbeat was stronger than all but three of the drivers' in the race."

Amazing. Not only is Carroll Shelby with us today because of a miracle of modern medicine, but the legacy of his world-beating Cobra continues in the form of the Viper. And that in itself is just as great a miracle. **MT**

VIPER VERSUS VETTE

VvV

DETHRONING THE KING OF THE HILL?

by C. Van Tune

PHOTOGRAPHY BY RICH COX

Until today, bragging rights belonged to Chevrolet. Its Corvette ZR-1 was undisputably the baddest animal in the domestic automotive jungle. A marauding lion among a highway full of quivering lambs. The top of the four-wheeled food chain. The true king of the hill.

As this is being written, however, production has begun on a new American sports car that'll give a severe case of angina to the "Heartbeat of America." For today, from within the walls of the New Mack assembly plant in downtown Detroit, a powerful contender to the throne has just been released: the Dodge Viper

RT/10. Created by the company famous (infamous?) for its compendium of K-car derivations, Viper takes all you know about Chrysler Corporation and snuffs it out with the first blat of its side-mounted exhaust. This is not only the quickest, best-handling car ever designed by the company, but also the most expensive. Yes, in a year when domestic car sales are among the worst in our lifetime, Chrysler has thrown sensibility to the wind and built a kick-tail tire-smoker for those of us in the lunatic inner circle.

Designed with an eye toward the past, Viper unabashedly emulates the Cobra 427 roadsters built by Carroll Shelby in the mid-'60s: Front engine, rear drive, and with as big a tire as possible stuffed into the wheelwells. Its 10-cylinder 8.0-liter (488-cubic-inch) megamotor boasts 400 horsepower at 4600 rpm and 450 pound-feet of torque at 3600 rpm. Quite simply, this is the largest engine ever used in a Dodge product, and eclipses even the powerful 426 Hemi of the muscle-car era. And while the Hemi handily won almost every championship from NASCAR to NHRA, Chrysler never put it into a two-seat sports car to compete with Corvette.

Therefore, it seems only natural to compare the (approximately) $50,000 Viper with today's Corvette ZR-1. Although at $65,318 the ZR-1 is significantly more expensive, the two cars have dead-even performance goals (0-60 mph in the fours; high 12-second quarter miles) and handling objectives (close to 1.0 g of lateral grip). Our first drives in two Viper development cars (*Motor Trend*, Nov. '91) gave us the incentive to proceed with this challenge. Will a head-to-head comparison dethrone the king, or is the flashy Dodge merely a show car with an inflated ego?

The opportunity to find out came recently during the press introduction of the Viper in Southern California. The activities included a 350-mile run along some of the best twisties in the state, interspersed with a couple of bouts of freeway driving, plus a half day of hot laps at Willow Springs Raceway. In short, a very good evaluation route.

Our original plan was to perform a complete road test of the Viper and ZR-1 along the driving route. Unfortunately, we were dealt a series of circumstances that limited our testing. First, '92 model ZR-1s weren't yet available, so we had to make do with a '91. There are no engine changes for the '92 ZR-1, but the effectiveness of the new Goodyear GS-C tires and Acceleration Slip Regulation (traction control) may account for some performance gain.

Second, all Vipers supplied for journalist evaluation were preproduction prototypes, which means items such as engine calibration, suspension settings, brake bias, and other minor features are subject to change before production begins. Thus, as with our past evaluations of preproduction vehicles, we'll be forced to give only initial impressions at this point and must wait until we get a production Viper before we can carve our test numbers, and our opinions, in stone.

At first glance of their spec charts, the Viper and Vette look similar. They share wheelbase dimensions of 96.2 inches and are only a few inches apart in overall length (178.6 inches, Vette; 175.1 inches, Viper). Both cars feature 17-inch wheels, massive four-wheel disc brakes, and six-speed manual transmissions. Each carries only two people, utilizes fiberglass-type body structures, and has a curb weight within 200 pounds of the other (3465, ZR-1; 3280, Viper).

That's where the similarities end. Merely park these machines side by side, and the Corvette practically fades away. Next to the Viper's Brahma-bull-on-steroids look, the conservatively styled ZR-1 appears as docile as a Corsica rent-a-car. More than appearances, a fundamental difference between the two can be found under their forward-tilting hoods. While Viper relies on the traditional American formula of a megacube motor to make its statement, the ZR-1 takes a more sophisticated approach (see sidebar). With its output of 375 horsepower, the ZR-1 boasts a weight-to-power ratio of 9.2 pounds/horsepower. The Viper's 400 horsepower equates to a ratio of 8.2 pounds/horsepower. Both cars utilize six-speed gearboxes (with first-to-fourth gear lockouts during light acceleration); the Corvette's is by ZF and Viper's is new from Borg-

Warner. It's interesting that internal gearing of both the trannies is nigh identical, with high 0.50:1 overdrives for each car's sixth gear. The heavier Corvette wisely chooses a 3.45:1 ring-and-pinion, while Viper makes do with tall 3.07:1 cogs. Both cars use limited-slip rearends.

As a result, sixth gear in either machine is practically useless at speeds under 80 mph. The ZR-1 is turning a mere 1500 rpm in that gear at 60 mph, while the Viper is loafing at a near-idle speed of 1200 rpm at the same velocity. The Dodge's first gear is good until 63 mph, and 100 mph in sixth equates to only 2000 rpm, which gives the car a theoretical top speed of 300 mph. Actual top speed is officially listed as 165 mph, but it's no secret Emerson Fittipaldi has seen the far side of 180 mph in a Viper at Chrysler's proving grounds.

VvV

We didn't get the chance to test top speeds, but were able to click off some unofficial acceleration runs. On an asphalt surface

THE POWER AND THE GLORY

This is an impossible assignment. How can you compare the engines in a Vette and Viper? Can you compare an F-15 Eagle with a P-51 Mustang, Madonna with Julie Andrews? Of course not. The Viper and ZR-1 engines produce roughly the same power, but the only other similarity is that they both burn gasoline. These two engines have personalities as diverse as Michael Jackson and Arnold Schwarzenegger (you guess which is which).

A comparison of specific outputs favors the Corvette LT-5 with 66 horsepower/liter against 50 horsepower/liter for the Viper. The LT-5 makes its power using the best of modern engine technology; the Viper uses brute displacement. No tricks like the Corvette's computer-controlled dual intake tracts—that's the domain of little (if you can call 5.7 liters little) high-revving engines. The 8.0-liter Viper is the new kid on the block and is based on a not-yet-in-production Dodge truck engine. Carroll Shelby and crew were quick to recognize the hype value of having the only production V-10 in town, and assigned corporate brother Lamborghini the job of turning this truck motor into a thoroughbred sports-car powerplant.

In truck form, the V-10 is all cast iron, but Lamborghini engineers re-cast the heads and block in aluminum, cutting 100 pounds from the iron engine. Lambo engineers were concerned with providing adequate cooling flow, so they designed an external water manifold to distribute coolant to individual cylinders and cylinder heads. The engine uses a single cam in the block, with hydraulic lifters and short pushrods operating two valves per cylinder. Compared to the truck cylinder heads, the Viper's valves are smaller and are placed farther away from the sides of the combustion chamber. A wedge was also cast between the valves to increase

No gratuitous high-technology here, just plenty of good old cubic inches and enough neck-snapping potential to bring a tear to the eye of any true muscle-car fanatic. The Viper's violent 8.0-liter OHV V-10 makes 400 horsepower at 4600 rpm and develops an awesome 450 pound-feet of torque at 3600 rpm.

with less than optimum traction, the Viper thundered to 60 mph in 4.8 seconds, went 0-100 mph in 12.0 seconds, and covered the quarter mile in 13.1 seconds with a terminal speed just under 110 mph. Fast, yes, but not the Cobra-beating figures we'd expected. In fact, the ZR-1 was closer to Viper than we'd imagined, with its times of 4.9, 11.8, and 13.4 seconds at 106.9 mph, respectively.

VvV

As important as Viper's all-out performance is its muscular torque curve. Whereas the ZR-1 needs to be in the correct gear for *acceleratus rapidimus*, the Viper driver needn't keep a close eye on which gear he's in. The big Dodge will leave the line in third or fourth gear without protest and deliver seamless power right on up to the 6000-rpm limit. The motor sounds healthy when you're leaning hard on the gas, but at idle or under light throttle, it's nearly hush quiet. By contrast, the Corvette's LT-5 engine delivers a high-rpm serenade better

turbulence. This design allows a 9.1:1 compression ratio without using knock sensors, uses premium unleaded, and meets emissions without needing EGR. The crankshaft was forged (cast in the truck), and forged truck connecting rods are used unchanged.

The Dodge V-10 uses a bottom-feed arrangement that places fuel injectors on the bottom of the intake plenum, controlled by a sequential EFI system, which is the first application of this technology at ChryCo. Bottom-mounted injectors lower the engine profile, and the fuel flow helps cool the injectors.

Stereolithography was used to computer-model the headers (particularly difficult to fit around the Viper's tube frame), saving the project an estimated six weeks over more conventional methods. The bottom-feed injectors and external water system resulted in the lowest profile of any 90-degree V-engine ever built at Chrysler, 25.9 inches from the bottom of the oil pan to the highest bracket on the motor.

The LT-5 Corvette motor uses almost every high-tech trick in the book to make 375 horsepower at 5800 rpm. It's at the cutting edge of engine technology, although it doesn't use any exotic metals or composites. Four cams are used, and four valves per cylinder. Two intake runners feed each cylinder, one per intake valve. Each intake valve has different valve timing, one biased toward high engine speed performance, the other low and medium speed. The engine control computer keeps the high-speed intake tract closed until high output is demanded by the driver, at which time it opens both tracts for true four-valve operation.

Mechanically, these two engines are dramatically different. From a power output standpoint, the Viper makes 25 more horsepower 1200 rpm earlier in the rev band than the Vette. The Viper's biggest real-world advantage is torque. This monster grunts out 450 pound-feet of torque at 3600 rpm (80 more than the Vette). The ZR-1 engine is the Joe Montana of sports cars—it's quick, intelligent, and knows all the tricks. The Viper's big V-10 is a massive defensive lineman—overpowering, always there, and above all, strong.

—Ron Grable

With its double overhead camshafts, four valves per cylinder, and dual-tract intake manifold, the all-alloy LT-5 V-8 is the antithesis of the Viper's monster V-10. Although it gives away considerable displacement, the 5.7-liter techno tour de force still makes a healthy 375 horsepower at 5800 rpm and 370 pound-feet of torque at 4800 rpm.

than anything found on a CD.

Because of strict government noise standards, Chrysler had to jump through hoops to get the side exhausts certified. The result is a set of stainless steel tubular headers that feed into close-coupled 12-inch catalytic converters and 21-inch mufflers. The entire piping has been outfitted with Nomex wrapping to reflect heat away from the body panels, but it still gets toasty along the door sills after you've been driving. A placard on each sill warns where not to touch.

VvV

In Europe of the '50s and '60s, a popular test of fast cars was 0-100-0 mph, and the Cobra 427 reputedly accomplished this synapse-abusing feat in a blistering 13.8 seconds. Like the Cobra, both Viper and Vette utilize excellent four-wheel disc braking systems, but Viper goes for brawn, not brains, with Brembo four-piston front calipers and 13-inch rotors all-around. Anti-lock is not used. The high-tech ZR-1 chooses Bosch ABS IIS anti-lock (augmenting 13-inch front rotors and 12-inch rears) and relies on its computerized sophistication to make up for the difference in swept area.

Even with its lack of ABS, the Viper easily bested the ZR-1 in this feat with a time of 15.8 seconds. The Corvette trailed with a 17.0-second run, but the numbers of both cars were significantly hampered by the road surface we used. On a proper test track, we're confident the Viper would come close to Chrysler's target of 14.5 seconds. We're not convinced a car this powerful should have to live without ABS, though, and understand it's being developed for future use.

When the Cobra was built, Shelby didn't have the luxury of equipping it with 17-inch wheels and tires (see sidebar). The Corvette uses 9.5-inch wide alloys and 275/40ZR17s in the front and 11.0-inchers with 315/35ZR17s at the rear. Viper fits the same size rubber on 0.5-inch wider front wheels, but takes traction to the extreme with 335/35ZR17

A LOOK AT THE LEGEND

Everyone from Chrysler President Bob Lutz to "Team Viper" consultant Carroll Shelby will tell you Viper was designed to be a modern-day "remembrance" of the brutal Cobra 427—an aluminum-bodied race car that just barely qualified as street legal. The limited run (348) of these monsters built between 1965-67 were truly not for the timid.

Shelby's goal was a simple one: create the meanest, nastiest, and plain old fastest production car on the planet. Few would dare question his ultimate success.

The first Cobra debuted in 1962 as an Anglo-American hybrid that combined a lightweight body and chassis from AC Cars of England with a powerful 260-cubic-inch V-8 from Ford. Carroll Shelby was the master chef of this international performance recipe, and served up a Ferrari-taunting dish that took the automotive press by the neck and shook them fiercely. The Cobra was suddenly on the cover of every major car magazine, and surprised even Shelby with its response. This was near the beginning of the American muscle-car era, and the 2100-pound Cobra was a radical notion among the 409-cubic-inch Chevy Biscaynes and big-by-comparison Corvettes that served as hi-po cars of the day.

Shelby didn't waste any time waiting to race his creation. In October 1962, Bill Krause drove the Cobra in its first event at the Times Grand Prix at Riverside Raceway. A broken rear hub sidelined the Cobra, but not before it ran rings around the Corvettes Shelby so hoped to beat. The 260 V-8 soon made way for the Ford 289, and performance jumped to supercar levels. The 0-60-mph run took a mere 5.6 seconds, and the quarter mile blurred your vision in the low 14s. (Remember, this was done on tires about as wide as those on your power mower.)

Shelby fought the Cobra's competition teething problems and continued to race, race, race. After defeats at Daytona, Sebring, and other venues, the men of Shelby-American hit on a formula that worked. Drivers with names like Bondurant, Gurney, Hill, and Miles took Cobras to victory across the country. In 1964, a special Cobra chassis was given an aerodynamic coupe body and sent forth to challenge Ferrari on its home turf. At LeMans, this Cobra "Daytona" scored a fourth overall finish and a first in GT class. In 1965, Shelby's cars won the coveted FIA GT class championship. The Cobras seemed unstoppable.

Street versions of the original Cobra roadster were stretched and formed to accept a strengthened chassis that was to contain the now-legendary 427-cubic-inch V-8. With a conservatively rated 425 horsepower (and 480 pound-feet) under its aluminum hood, the 2529-pound Cobra 427 was untouchable in straight-line performance. But with its 90.0-inch wheelbase and 48/52-percent front-rear weight distribution, this Cobra was a four-wheeled frag bomb in the hands of anyone but the most talented drivers. The car pictured here, owned by Ron McClure, yanked our fifth wheel to 60 mph in 4.2 seconds and through the quarter mile in 12.8 seconds at 111.3 mph. But give the Viper a set of 4.11:1 rear gears, and it'd be just as quick. Who says they don't build 'em like they used to? —*CVT*

TECH DATA

'66 Cobra 427

GENERAL/POWERTRAIN	
Body style	2-door, 2-passenger
Vehicle configuration	Front engine, rear drive
Engine configuration	V-8, OHV, 2 valves/cylinder
Engine displacement, ci/cc	427/6998
Horsepower, hp @ rpm, SAE net	425 @ 6000
Torque, lb-ft @ rpm, SAE net	480 @ 3700
Transmission	4-speed man.
Axle ratio	3.54:1
DIMENSIONS	
Wheelbase, in./mm	90.0/2286
Length, in./mm	156.0/3962
Curb weight, lb	2529
Fuel capacity, gal	18.0
CHASSIS	
Suspension, f/r	Independent/independent
Steering	Rack and pinion
Brakes, f/r	Vented discs/vented discs
Wheels	15 x 7.5, cast magnesium
Tires	8.15 x 15 Goodyear Blue Dot
PERFORMANCE	
Acceleration, 0-60, sec	4.2
Quarter mile, sec/mph	12.8/111.3
Braking, 60-0, ft	130
Skidpad, 200-ft, lateral g	0.92 (est.)
PRICE	
Base price	$7495
Typical restored price (1992)	$500,000

radials on 13-inch-wide rear alloys.

Suspension development has also made big leaps since 1965. Like Cobra, Viper uses a tubular steel frame, but includes a center spine structure with tubular outriggers to support the resin transfer molding composite body panels. High torsional stiffness was a major design criteria, and the result is a car that shows remarkably little flex when driven over railroad tracks or potholes. The suspension is fully independent, with unequal-length upper and lower control arms backed by coil-over Koni gas-charged shocks. Tubular anti-roll bars (27-millimeter diameter front, 22 millimeter rear) and a power-assisted rack-and-pinion box (16.7:1 ratio) contribute to the car's fast, flat handling. The 630-pound V-10 is mounted 6.5 inches behind the centerline of the front axle to allow a perfect 50/50 front/rear weight distribution.

The Corvette uses its tried-and-true combination of upper and lower control arms plus a transversely mounted fiberglass spring at the front. Upper and lower control arms, trailing links, and a similar fiberglass spring at the rear provide quick response to driver inputs and delights with an ultra-flat cornering attitude. Cockpit-adjustable Delco-Bilstein gas-charged shocks and anti-roll bars serve duty on both ends of the suspension. Power rack-and-pinion steering with a 15.7:1 ratio is also employed. Weight distribution is a slightly front-heavy 52/48 percent.

Comparing the cars on the street, through curving mountain roads, and across flat-out rural landscapes rewarded both our drivers with permanent grins. But back-to-back stints behind each car's leather-wrapped steering wheel brought out some big differences.

The Corvette is a perennial favorite of ours, with a winning combination of ready power and excellent

driver feedback. This is a machine you can hang out in a turn with confidence, and it can be steered with the throttle in all but the really tight, slow, corners that evoke understeer. The FX-3 adjustable suspension allows tuning for sport or comfort, and the big tires provide ample warning before breaking loose. On the down side, the ultimate Corvette suffers from a transmission that sounds like a cement mixer and an engine that tends to run quite hot when exercised hard.

Jump into the Vette after a half-hour blast (and we mean blast!) in the Viper, and the ZR-1 feels like a pussycat by comparison. Neither car can be characterized as difficult to drive, but the Viper will bite an untalented driver if its throttle is lifted abruptly in a fast corner. Steering effort is heavy, but with good feel on-center. Response is fast and linear without being darty. There's effectively no body roll, and the superior frame rigidity makes it feel as solid as a bank vault.

VvV

On slow corners, an initial bit of understeer warns you're entering too fast, but kick it gently with the throttle, and the tail will walk around to correct the line. Driven with finesse, we're convinced the Viper can see 1.0 g on a skidpad, which would best the top ZR-1 rating of 0.95 g by a significant margin. Not a high-strung exotic, the Viper can be chugged along in 10-mph traffic without complaint, and average drivers can have fun in the car as long as they have respect for the throttle and the fury it controls.

The Vette's cockpit is Ritz Carlton plush compared to the Viper's no-power-anything attitude. There are no exterior door handles or side win-

dow glass in this Dodge, and you'd better get used to the rushing wind, as heavy turbulence around the B-pillars will whisk your ballcap away at speeds over 90 mph. A removable canvas "toupee" top (weighing 16 pounds) is standard and features snap-in plastic side curtains. We tested the top (without side curtains) in a brief run up to 140 mph and were impressed with its tight fit and ability to dramatically reduce turbulence.

The Viper's driving position is more upright than the Corvette's, and we much prefer the Dodge's no-nonsense display of analog gauges. Three-point door-mounted belts (no airbag) are used, while Corvette owners receive conventional belts and an airbag; a Viper option well worth buying is a set of five-point harnesses. Chrysler's minimalist-thinking engineers acquiesced only to allow leather bucket seats with pump-up lumbar supports, tilt wheel, and powerful six-speaker sound system as concessions to anything but road-race use. On the other end on the sybaritic spectrum, the ZR-1 pampers with power assist on everything but the glovebox door. To appease the landed gentry, air conditioning will be available on the Viper, but only as a dealer-installed option.

There's ample leg room in the Viper for people over six feet tall, but the pedals are offset far to the left and force the driver's clutch leg uncomfortably against the bottom of the dash. Ergonomically, the Viper's shifter is easier to manipulate, but the Corvette's parking brake is handier to use. Forward vision is better in the Dodge, but its rear view is worse due to the bulging back fenders that occlude much of the mirror's field of vision. Additionally, the Viper's "targa bar" and short rear glass render the center mirror practically useless. Like Cobra, Viper includes a real trunk with space for two golfbags.

So both cars are brutally fast, give nosebleed handling, and are as exciting (almost) to strap on as an Indy car. But does the Dodge Viper RT/10 indeed dethrone the "king of the hill" Corvette ZR-1? Many people (author included) feel the Viper is a standout performance bargain and far surpasses the ZR-1 in thrills per mile. Others point out that the Corvette can be driven come rain, snow, or shine, and give it points for comfort and technology. But when the discussion's over and it's time to go drive, there's no waffling: The Viper is our candidate for 1992 king of the hill. Long live the King. MT

TECH DATA

GENERAL

	Chevrolet Corvette ZR-1	Dodge Viper RT/10
Make and model	Chevrolet Corvette ZR-1	Dodge Viper RT/10
Manufacturer	Chevrolet div., General Motors Corp., Warren, Mich.	Chrysler Motor Corp., Highland Park, Mich.
Body style	2-door, 2-passenger	2-door, 2-passenger
Drivetrain layout	Front engine, rear drive	Front engine, rear drive
Base price	$65,318	$50,000 (est.)

DIMENSIONS

	Chevrolet Corvette ZR-1	Dodge Viper RT/10
Wheelbase, in./mm	96.2/2444	96.2/2444
Track, f/r, in./mm	59.6/60.4/1513/1534	59.6/60.6/1514/1539
Length, in./mm	178.6/4536	175.1/4448
Width, in./mm	74.0/1880	75.7/1923
Height, in./mm	46.7/1186	Not available
Ground clearance, in./mm	4.7/120	Not available
Weight distribution, f/r, %	52/48	50/50
Manufacturer's curb weight, lb	3465	3280
Weight/power ratio, lb/hp	9.2	8.2
Fuel capacity, gal	20.0	22.0

ENGINE

	Chevrolet Corvette ZR-1	Dodge Viper RT/10
Type	V-8, liquid cooled, cast aluminumn block and heads	V-10, liquid cooled, cast aluminum block and heads
Bore x stroke, in./mm	3.90 x 3.66/99.0 x 93.0	4.00 x 3.88/101.6 x 98.6
Displacement, ci/cc	349/5727	488/7990
Compression ratio	11.25:1	9.1:1
Valve gear	DOHC, 4 valves/cylinder	OHV, 2 valves/cylinder
Horsepower, hp @ rpm, SAE net	375 @ 5800	400 @ 4600
Torque, lb-ft @ rpm, SAE net	370 @ 4800	450 @ 3600
Horsepower/liter	66	50.1
Redline, rpm	7000	6000
Recommended fuel	Unleaded premium	Unleaded premium

DRIVELINE

	Chevrolet Corvette ZR-1	Dodge Viper RT/10
Transmission type	6-speed man.	6-speed man.
Gear ratios		
(1st)	2.68:1	2.66:1
(2nd)	1.80:1	1.78:1
(3rd)	1.29:1	1.30:1
(4th)	1.00:1	1.00:1
(5th)	0.75:1	0.74:1
(6th)	0.50:1	0.50:1
Axle ratio	3.45:1	3.07:1
Final-drive ratio	1.73:1	1.54:1
Engine rpm, 60 mph in top gear	1500	1200

CHASSIS

	Chevrolet Corvette ZR-1	Dodge Viper RT/10
Suspension	Upper and lower control arms, monoleaf springs, anti-roll bar	Upper and lower control arms, coil springs, anti-roll bar
Brakes, f/r	Vented discs/vented discs	Vented discs/vented discs
Anti-lock	Standard	Not offered
Steering	Rack and pinion, power assist	Rack and pinion, power assist
Wheel size, f/r, in.	17 x 9.5/17 x 11.0	17 x 10.0/17 x 13.0
Wheel type/material	Cast aluminum	Cast aluminum
Tire size, f/r	275/40ZR17/315/35ZR17	275/40ZR17/335/35ZR17
Tire mfr. and model	Goodyear Eagle ZR	Michelin XGT Z

PERFORMANCE AND TEST DATA

	Chevrolet Corvette ZR-1	Dodge Viper RT/10
Acceleration, sec		
0-30 mph	2.1	2.0*
0-40 mph	3.0	2.8*
0-50 mph	3.9	3.9*
0-60 mph	4.9	4.8*
0-70 mph	6.4	6.1*
0-80 mph	7.6	8.4*
0-90 mph	9.7	10.2*
0-100 mph	11.8	12.0*
Standing quarter mile, sec @ mph	13.4/106.9	13.1/109.3*
Braking, ft		
60-0 mph	120	138*

**Note: Viper performance figures are unofficial.*

FUEL ECONOMY

	Chevrolet Corvette ZR-1	Dodge Viper RT/10
EPA, city/hwy., mpg	16/25	Not available
Est. range, city/hwy., miles	320/500	Not available

DODGE VIPER

Standing one metre away, you feel the heavy metal breathing of its 300 kW, eight-litre V10. Jukka Sihvonen was one of the first journalists to get his hands on the car Bob Lutz described as 'an exciting, gut-wrenching, out-and-out sports car with no frills'...

Touted by its Chrysler creators as the 'nineties successor to the Cobra, the Viper exudes power rather than elegance: a 44 magnum styling statement that makes no apologies to Pininfarina.

STANDING one metre from the Dodge Viper, one can feel the heavy-metal breathing of its 300 kilowatt, eight-litre V10. The driver nudges the throttle and the exhaust rushing from the huge sidepipes whumps you, just about crotch level.

That is wonderfully appropriate. After all, that is precisely where Dodge has aimed the Viper. This is not a cerebral, high-tech engineering exercise. The Viper is old-time religion – a visceral experience at modern-day prices.

The original Viper made its debut (as a concept vehicle) at the North American Automobile Show in 1989. Since then, the Viper Gang of Three – Chrysler poohbah Bob Lutz, Chrysler engineering director Francois Castaing and Chrysler chief stylist Tom Gale – have been Viper-active, enthusiastically touting the two-seater as the 'nineties successor to the Cobra.

"Sports cars have gotten way too East Coast, effete... effeminate, with lots of whizzy little parts in there and way too much electronics and too much sophistication," declared Lutz in 1989. The Viper he envisioned would be a "true sports car... an exciting, gut-wrenching, basic automobile with no frills... a connection of man and machine with a minimum of complication."

Using an 85-strong team including engineers, designers, manufacturing, finance and purchasing agents and employing simultaneous engineering, the Viper is being brought to market in what is for Chrysler record time – just under 36 months. The development costs are estimated at 20 million US dollars, with another 50 million set aside for tooling and manufacturing costs.

Some Chrysler executives, including Lutz, believe that the fact that more than a year was cut off Chrysler's normal development time may be the most significant thing about the Viper. The car is now scheduled to go on sale in the United States in early 1992 with about 200 cars built the first year.

After that, volume may increase to as much as 3 000 a year and there are plans to sell cars in Europe early in 1993. "What we want is to go and challenge the Germans on their home ground," says Jean Mallebay Vacquer, Chrysler's general manager of special projects' engineering.

A pair of prototypes

On a fine, warm day at Chrysler's proving grounds in Michigan, after two years of public relations foreplay, it is finally time for the first non-Chrysler employees to drive a pair of Viper prototypes. One had been driven by ex-racer and Viper enthusiast Carroll Shelby as the pace car at the 1991 Indianapolis 500. The other is a slightly ragged development car. They are worth about 300 000 dollars each. There is still some fine-tuning to be done, but Chrysler officials insist the pair are very close to what production models will be like.

The Viper's appearance has changed little from the 1989 concept car. It is still attention getting, purposeful and anything but delicate or understated.

This is no quietly elegant Pininfarina design. This is a 44 magnum styling statement that passes "strong" and moves into "blatant", further accentuated by a somewhat silly decal of a snake on the bonnet. One could drive forever in a ZR1 Corvette and only the cognoscenti would notice. But even Stevie Wonder would never miss a Viper.

The overall length is 4 369 mm, arranged over a 2 438 mm wheelbase. The Viper, which will be built in Detroit, has a rectangular tube frame with plastic composite body panels formed with low-pressure moulding and affixed with mechanical fasteners.

The interior is as basic as the concept. Air-conditioning won't even be available, despite a price described by straight-faced Chrysler spokesmen as "hopefully under 50 000 dollars." The unspoken attitude suggests that the Viper is a driver's car and creature-comfort seeking sissies should spend their disposable income elsewhere.

The tachometer and speedometer are large. Black on white. Simple, with the speedometer reading 180 mph. Glance to the right a bit and there's information on coolant temperature, oil pressure, fuel level and volts.

There is an AM/FM radio/cassette, with the speakers located between the seats, facing forward. Tweeters are located in the doors. There is a snap-in soft top that can be stored in the smallish boot, but as of now it had yet to be tested for high-speed "viability".

There is no airbag. Instead, Chrysler will require Viper-isti to clamber beneath a door-mounted "passive" system required to meet federal safety regulations. However, it plans to provide mounting hardware for those who would like five-point racing harnesses.

The seats are comfortable, but adjustments are limited to fore-and-aft and seatback rake.

The foot room is also tight, with nearness of the wheel meaning there is no dead pedal. That leaves tall drivers wondering where to store the left foot when it is not pushing up and down, driving what Chrysler hopes will be the Lord of the Flyers. The good news is that the pedals are nicely aligned, for those of the heel-and-toe persuasion.

V10 'wants out'...

Even at idle, the V10 shakes the Viper. Lots. It is as if something really big and not particularly refined is trapped under the hood and is moderately serious about wanting out. This engine, after all, is based on a V10 truck engine Chrysler plans to introduce on a new pick-up. But instead of being annoying, the vibro-massage seems somehow in keeping with the Viper's persona.

The clutch take-up is sharp, the pressure moderately heavy but not so severe that it will substitute for Nautilus. Not surprisingly, the Viper's acceleration is immensely adequate and equally immediate.

Chrysler claims the V10, with its

9,5:1 compression and multi-point fuel injection, produces 300 kW at 4 600 r/min and 610 N.m of torque at 3 200 r/min. It uses a 3,07:1 Dana final drive. The current kerb weight is about 1 454 kg. That is 91 kg over the target weight and Dodge plans to introduce the Viper at this figure and trim some mass off later.

Chrysler was not allowing our small group any "instrumented testing". But they claim that even at its overly chubby weight, the Viper meets its goals.

They are: zero-to-60 mph (96,6 km/h) in four seconds, zero-to-100 mph (161 km/h) and back to zero in 14,5 seconds, top speed of 257 km/h, a quarter mile at 12,7 seconds and pulling 1,0 Gs on a 91,4 m skid pad.

The six-speed gearbox is a Borg-Warner design, chosen after a Getrag unit was dumped over what Chrysler said were "economic issues". The pattern is identical to a conventional five-speed, with sixth straight down and reverse available up and to the right of fifth.

The shift lever is light and a little notchy, but quite acceptable. But with so much torque, those who shun manual labour can almost use the six-speed as an automatic.

After all, Chrysler claims the all-aluminium, 20-valve V10 is already producing 542 N.m of torque at 1 600 r/min and my driving impressions make that assessment seem quite reasonable.

Get hard on the throttle, which is how the designers intend the Viper to be driven, and any opportunity for conversation ends, eliminated by an engine/exhaust duet. It is a deep noise, distinctive but not sounding as raucously sophisticated as a Porsche 911.

Under full throttle acceleration, Viper Sound can easily be heard 400 metres away. That is one of the problems that worries Viper engineers, when they prepare to market the car in Europe. The acceptable limits are low.

"Switzerland is our worst, 75 decibels," says one Viper engineer during a staff meeting later that day. "75 decibels? Can we shut the engine off and coast through (the test)?" responds Chrysler powertrain expert Chris Theodore.

Another major change for Euro-bound Vipers involves the side exhaust pipes. The American version used pressure-formed Nomex with stainless steel cladding to reduce the chances of the unwary being burned. Chrysler believes such sidepipes would be illegal in Europe, so the exhaust will be routed traditionally.

With a V10 hunkered down up front, one expects the Viper to be nose-heavy. Wrong. Chrysler has adopted a front mid-ship design that pushes the V10 165 mm behind the axle.

That requires the pedals to be skewed slightly to the left, but the design provides a 50/50 weight distribution. Particularly for its size, the Viper is remarkably eager to change directions offering a surprisingly sharp turn-in. The steering is tight, if slightly uncommunicative. It is willing to obey, but no

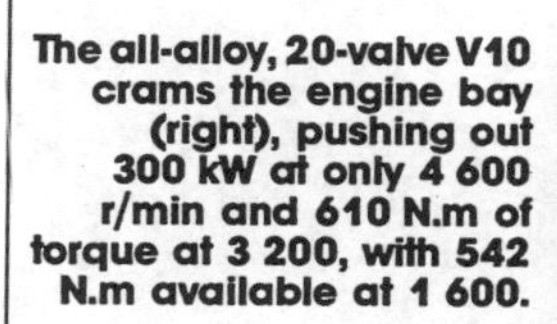

The all-alloy, 20-valve V10 crams the engine bay (right), pushing out 300 kW at only 4 600 r/min and 610 N.m of torque at 3 200, with 542 N.m available at 1 600.

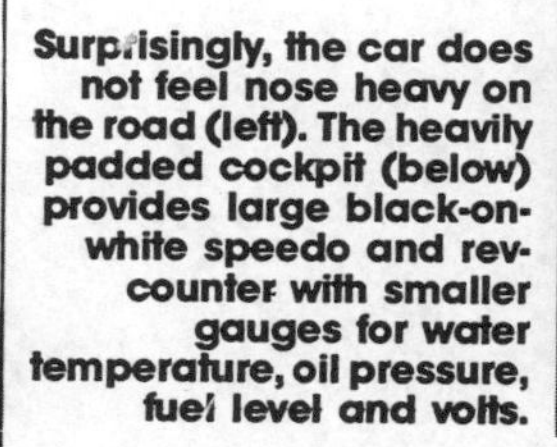

Surprisingly, the car does not feel nose heavy on the road (left). The heavily padded cockpit (below) provides large black-on-white speedo and rev-counter with smaller gauges for water temperature, oil pressure, fuel level and volts.

chatty.

There was a slight but noticeable difference in steering feel between the two vehicles, the result of on-going development work. Some drivers felt one was slightly loose. The vehicle I drove, however, was fine.

The suspension is fully independent, using short-long arms, coil springs over Koni shocks and front and rear stabiliser bars. Through a cone-marked road course, the Viper's body motions were tightly controlled with minimal body lean and the structural rigidity felt good. On the smooth surfaces Chrysler chose for our drive however, both ride quality and scuttle shake remained question marks.

Although one hesitates to flail about excessively in one of two such expensive prototypes, the grip seemed impressive and the Viper is likely to be yet another sports car whose capabilities wildly exceed those of most of its drivers. Part of that adhesion is thanks to God's little acre of rubber underneath. The Viper uses Michelin's XGT-Zs with 275/40/17s up front and 335/35/17s in the rear.

Move into a left-hand sweeper at about 130 km/h and back off the throttle. The tail moves slightly wide but the movement is quite benign. Helpful, not threatening. The nose edges a little deeper into the turn but understeer still dominates. "Because we can't control who is going to buy it, we are trying to develop a vehicle that is not going to get the driver in trouble," explains Viper engineer Herb Helbig.

One driver complains that the other prototype, the one he drives, is light and twitchy at the tail. But that is not the case with ours.

A tight, decreasing radius comes up. The Viper's brakes are 330 mm ventilated discs, front and rear, with the fronts using a four-piston caliper. The pedal brake has a take-no-prisoners firmness.

It feels like serious stopping power, but none of that new fangled stuff from Teves or Bosch. "We have no anti-lock, just big brakes," says one Viper engineer.

At the engineering meeting later, Chrysler engineers will discuss the hard pads and the probability that at some time, the Viper's brakes will simply and rudely squeal. Rather than change the brakes, they are seriously considering putting a notice in the owner's manual that the Viper is a performance car and the noise is normal. Take two aspirins and don't call your dealer in the morning...

A few more laps and it is time to get out of the Viper. Is America's next sports car going to match up to the Viper-hype? Can one justify a 50 000 dollar tag for a sports car without increasingly common technology such as all-wheel drive, rear-wheel steer, traction control, an adaptive suspension or even anti-lock brakes?

The answer will take more quality time behind the wheel and with finished vehicles, not prototypes. But finding out is going to be fun. ●

SNAKE BITTEN!

Dodge's Viper RT/10 strikes, with all the performance venom of the legendary Cobra

BY RON SESSIONS

HOLLYWOOD, CALIFORNIA— cruising Sunset Boulevard top down in the sensational, new Dodge Viper RT/10. The late afternoon sun has dipped behind the mountains on this surprisingly balmy November day. On the sidewalks, an eclectic mix of recording-industry executives, tourists, boutique-store operators and street people are stirring. Out over the Viper's long, curvaceous hood, an endless row of red stoplights stretches out before me as rush-hour traffic clogs the twilight. I'm inching along in 1st gear now, and it gives me pause to reflect.

It has been two days since I first keyed the Viper to life. Some 300 miles later, a driving route that included freeway cruising, delicious cut-and-thrust twisty bits, wicked mountain switchbacks, wide-open stretches of high desert and some apex-clipping hot laps at Willow Springs raceway has afforded me a full measure of quality man-meets-machine bonding time. And with a tangled nest of split ends that passes for hair, I have the Viper-doo to prove it. Heady stuff, all.

But now, stuck in traffic, I have time to refocus on the cultural wake this roadster is creating. In Hollywood, where anything and everything goes, one would think that very little gets anyone's attention. A quick scan of the Viper perimeter and I can't help noticing; *people are staring.* Traversing this strip of surreal real estate in Dodge's V-10 wonder, I feel about as discreet as Hannibal sacking Rome.

The Viper turns heads.

It elicits whoops and hollers of approval. A pair of Young Turks in a Toyota MR2 let loose with a lusty catcall of the sort usually reserved for the L.A. Lakers cheerleaders. A middle-aged couple in a Bronco with Indiana plates pull up alongside and flash the thumbs-up sign. A biker who could pass for one of the Grateful Dead chugs his Harley even with the Viper to cut a gap-toothed smile and a nod of recognition. "Dig it man; the Harley Hog of sports cars."

When conversation is possible, peo-

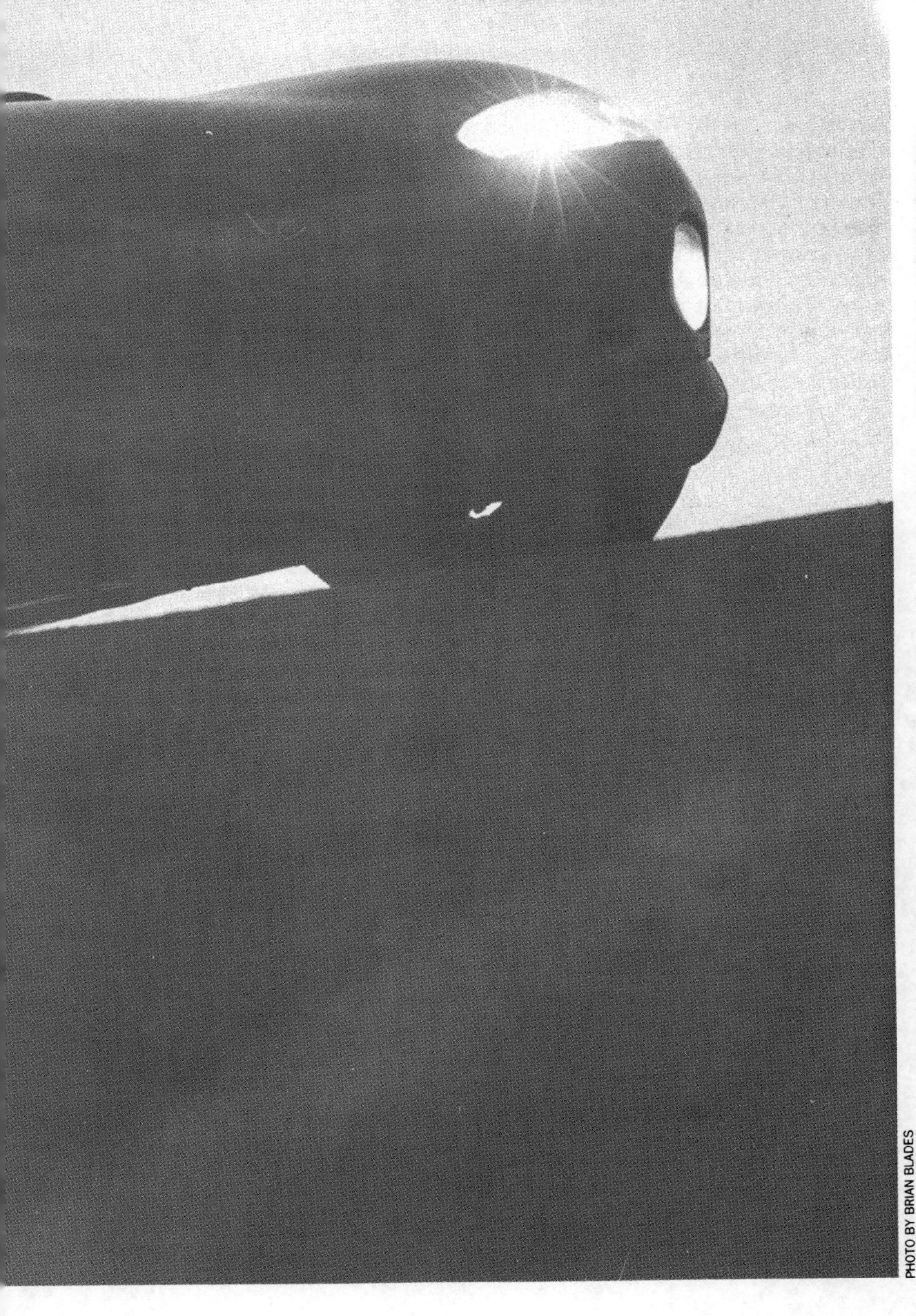
PHOTO BY BRIAN BLADES

ple in cars or on the street query: "So, how's the Viper?" or "Iacocca actually pulled it off, didn't he?" No confusing this wonderfully outrageous red 2-seater with some VW-powered kit car or high-dollar techno-marvel. When is the last time an American car, let alone one from Chrysler, has caused such a stir?

Why?

For starters, the Viper's visage is imposing. A menacing wide-mouth grille and animalistic polyellipsoid headlamps look as though they just made a long journey up the River Styx. Hints of the Viper's inspirational forebear, the Shelby Cobra 427, are recognizable: in the curved haunches of the rear fenders, in the basic honesty of the instrument panel, in the muscular bulges of the hood and front fenders.

The Viper sits low and wide, as if ready to uncoil with explosive force at a moment's notice. At 75.7 in., it's wider than a Corvette ZR-1 or Ferrari 348. Standing just 44.0 in. high, the Viper squats lower than a Ferrari F40 or Testarossa and more than 2½ in. lower than the Acura NSX.

Reptilian gill-like vents at the rear edges of the front fenders give a hint of the massive V-10 engine that lies beneath the Viper's hood, and they're functional heat-exchangers too. The crowning bad-to-the-bone touch is the side-mounted exhaust pipes, the first such devices standard on a production car in 25 years (Vipers sold in the state of Connecticut and 1993 models exported to Europe will have a rear-exit exhaust still under development because sidepipes are illegal there).

A targalike structural bar hints of the roadster version of the Ford GT40. Flying in the face of more sophisticated exotics with their climate-controlled cockpits, the Viper sports an open-air nature that validates the notion that the pleasures of driving it are of this Earth.

Light on the accouterments, heavy on the performance hardware, the Viper is about as different from traditional Chrysler fare (K-Cars, minivans, padded-vinyl-roof New Yorkers and Jeeps) as one could imagine. A 2-seat open sports car powered by an all-aluminum V-10 engine, with rear drive, a tube frame and a plastic body, sounds like something conjured up in Hethel, Maranello or Munich—certainly not the Motor City. Headed by a GM escapee, Executive Engineer Roy H. Sjoberg, Team Viper consists of 85 carefully screened "car nuts" from within Chrysler who volunteered to work on the project. Huddled in a warehouselike skunk works in a semi-seedy section of west Detroit (previously the old AMC Jeep/Truck Engineering building), the Viper development project represented something of a mini-Manhattan Project for Chrysler, rushing the roadster from concept car to dealer showrooms in a scant 36 months.

And though every dimension and body panel have been changed to meet federal safety regulations or production-line realities, the Viper remains true to the form of the original showstopper that wowed 'em on the Chrysler turntable at the 1989 North American International Auto Show in Detroit (see R&T, April 1989). If a committee has been at work here, it's not the proverbial camel-causing confab that Detroit's been infamous for. The Viper now has 5-mph bumpers, a legal-height windshield, passive seatbelts, side-exit exhausts that meet stringent noise guidelines, European-homologated lighting systems, huge 13.0-in. disc brakes and full emissions controls.

By the time you read this, the Viper will have begun production out of the small, New Mack Avenue facility in east Detroit (formerly a shop where Chrysler built prototypes). There, between 120 and 160 workers, dubbed "craftspersons" and organized into groups of five, will essentially hand-assemble Vipers with the help of computerized inspection equipment.

A scant 200 cars will be produced in

model year 1992. By 1993, planned production rises to 2000 cars, sales to Europe beginning as well. Ultimately, Chrysler says it can build 3000–5000 Vipers per year, but that may be overestimating the size of the open, 2-seat near-exotic niche. As a point of reference, Shelby built just under 400 Cobra 427s in two years, and Honda is having more difficulty than expected selling its annual U.S. allotment of 3000 Acura NSXs. Time will tell if a single-purpose car with no outside door handles, roll-up windows, automatic transmission or factory air conditioning can sell in the projected numbers. Initially, anyway, a two-year Dodge Viper waiting list and stratospheric dealer premiums are a near certainty.

But enough of history and conjecture. The Viper's appeal is unabashedly emotional. The adrenaline starts pumping and the revelations begin the moment you lay eyes upon this retro roadster.

PHOTOS BY JOHN LAMM

■ Minimalist to the max, the Viper's take-charge cockpit beckons the driver to the open road. Analog gauges with backlit markings come alive at night. Extra room behind the instrument panel accommodates dealer-installed air conditioning. Though you can't lock up a Viper, a keyfob-activated security system sounds the horn and deactivates the ignition if triggered. Message inside the glovebox epitomizes Team Viper's spirit.

As with the Cobra and numerous classic British roadsters, there are no outside door handles; you reach inside and with a backhand flip, pull open the inside door-release handle. That long black object brushing against your pant leg is the engine exhaust sidepipe, and just upstream of it is one of the car's two catalytic converters. If the car's been running recently, it's best to stay clear. A large, prominently placed sticker at the rear of each door opening exclaims, "WARNING: HOT EXHAUST PIPES BELOW DOOR OPENING—AVOID CONTACTING THIS AREA" (a 3-mm-thick layer of Nomex minimizes heat transfer into the cabin).

A short, easy hike over the sill and sidepipes puts you into a supportive, no-nonsense bucket seat that seems to fit drivers short or tall, large or small. Aside from the pipes, ingress and egress are much easier than in, say, a Corvette. There are but two adjustments, seatback rake and fore/aft—no others are necessary. The passive restraints are door-mounted belts with well-located outboard anchor points. Merely close the door and buckle the "passive" belt as you would in a Nissan NX or Chevy Lumina. The footwell is not cramped, but because the pedals are shifted to the left to clear the front midships engine configuration, there's no room remaining for a dead pedal.

Seated at the controls, a leather-wrapped, 3-spoke steering wheel of robust construction beckons you to the open road. There's a wonderfully simple yet attractive instrument layout with 7000-rpm tachometer, 180-mph speedometer and warning-lamp binnacles centered in front of the driver. Auxiliary gauges for coolant temperature, oil pressure, fuel level and volts trail off to the right. The Viper's gauges have a dual personality—gray faces with black pointers and lettering by day; backlit pointers and lettering in a vibrant yellow with red highlights at night. Rounding out the minimalist dash are a simple push/pull headlamp switch, foglamp toggles, rotary controls for heating and ventilation (but not air conditioning), and the one obvious concession to Sybaritic conduct, a 6-speaker AM/FM stereo/cassette, which Chrysler wags insist is enjoyable up to 100 mph. (Slide in a Steppenwolf tape, select "Born To Be Wild" and you're off.)

The Viper has a high driveline tunnel, which also houses structural members that help account for its outstanding 5000 lb.-ft. per degree torsional stiffness. Sprouting somewhat awkwardly out of the tunnel is the handbrake, borrowed from the LeBaron convertible, and the 6-speed shifter. The Viper's gearshift has relatively long throws, but easy-to-find, precise gates allow slam-dunk shifts or just puttering around town. What looks like tasteful gray, crackle-finish plastic on the dash, doors and tunnel is actually structural urethane-foam trim—the first on a U.S.-built car.

Another first on a modern production car is Viper's all-aluminum V-10 engine. As it should be, this 400-bhp, 8.0-liter (488 cu.-in.) powerplant is the Viper's heart and soul. Derived from a similar V-10 of the same displacement with cast-iron heads and block that's scheduled to appear in Dodge trucks in 1993, the Viper V-10's architecture is essentially that of Chrysler's small-block 5.9-liter (360-cu.-in.) V-8 with two cylinders grafted on.

As with the 427 Cobra, the Viper V-10's long suit is prodigious torque. Though it peaks with 450 lb.-ft. at 3600 rpm, the V-10's torque curve is a broad plateau extending from 1500 to 5500 rpm. A sequential multipoint fuel-injection system with bottom-fed injectors, dual throttle bodies and dual plenums provides excellent driveability, and tuned intake runners give a ram-tuning effect between 2000 and 4000 rpm.

Early in the Viper program, Lamborghini Engineering was brought in for its expertise with high-performance, aluminum-block engines. Among the numerous improvements Lamborghini made to the aluminum V-10 is a Formula 1-inspired external coolant manifold running alongside the block. As a result, the engine has the lowest coolant-temperature rise of any engine Chrysler has ever built. Casting the block and heads from aluminum also has a weight benefit, saving 100 lb. compared with the truck V-10. Also of interest from a materials standpoint are the Viper's magnesium valve covers and cast-steel tuned exhaust headers, a close fit within the Viper's frame rails.

Fire up the engine, and all of this specification talk fades away. Because of the uneven firing pulses (occurring at 90 and 54 degrees of crankshaft rotation), the exhaust note is sort of a macho staccato chugga-chugga with a slight wheeze. I grew up on MoPars in the Sixties so I can say this; on first blush, the Viper's song sounds a bit sour, like two Slant Six Plymouth Valiants, each down a cylinder, with pinholes in their mufflers. Commuting to college in a Valiant with a pinhole in its muffler, I thought it sounded cool (silly youth). The Viper's phonics won't remind you of Sebring 1966, or the NHRA Winternationals, or Talladega or even today's Bob's Big Boy on a Saturday night. A Street Hemi, it's not. But you get used to it. You even get conditioned in a sort of a positive Pavlovian response to the Viper's exhaust sound because of the zoomy things that happen as the sound gets louder.

Getting the Viper, mini-catalysts, Walker sidepipes and all, to pass noise

Wickedly fulfilling 7990-cc aluminum V-10 gives 165 mph, and with just 8.3 lb. per each of the Viper's 400 horses, 0–60 mph takes 4.5 sec. Robust construction includes forged-steel connecting rods, crankshaft.

regulations was no small task. The big V-10 breathes quite well right up to its 6000-rpm redline and the pipes keep the exhaust din below the federal 80-dBA threshold.

An all-new Borg-Warner T56 6-speed transmission and hydraulically actuated 12-in. clutch get the Viper in gear and down the road in a hurry. The Borg-Warner engineers claim to have paid special attention to geartooth micro-finish and gear spacing. The result is a truly modern, quiet, easy-shifting close-ratio transmission with two overdrive gears, not at all like the crash-boxes of yore. It's certainly quieter than the Corvette's ZF 6-speed. And like the other popular brand across town, the Viper's box has a computer-aided 1-to-4 shift to help with the EPA city-cycle fuel mileage rating (thereby minimizing the gas-guzzler tax). When puttering along in 1st gear in the 15–25-mph speed range with a warm engine and under steady part-throttle, the computer blocks the 1–2 gate and ensures that any shift you make is into 4th gear. Although the V-10 is dozing at idle speed at 20 mph in 4th, it doesn't protest a bit. No automatic transmission is offered.

Aft of the gearbox, a short aluminum driveshaft goes to a limited-slip 3.07:1 Dana 44 differential.

Fresh out of the box, Chrysler claims supercar straight-line performance numbers: 0–60-mph in 4.5 seconds and the quarter-mile in 12.9 sec. at 113 mph. I had the opportunity to hook up a Vericom performance computer for a half-dozen runs to validate my seat-of-the-pants observations. With photographer Brian Blades and all his gear aboard and running in 95-degree heat at 3500 ft. above sea level, I managed a 4.9-sec. best 0–60 and a 13.2-sec.-at-109-mph quarter-mile run. So the Chrysler estimates sound reasonable.

The Viper slithers forth wearing a composite skin formed largely by the resin transfer molding (RTM) process. Only the lower front body enclosure is formed of sheet-molded compound, which is also found on the Corvette. Both materials can be loosely described as fiberglass. The newer RTM process is used on such European sports cars as the Lotus Elan and Esprit, BMW Z1 and Alfa Romeo SZ.

Underneath the Viper's sinewy skin is a surprisingly rigid tubular steel backbone frame. Many open cars have all the structural integrity of an open shoebox, but the Viper's stout skeleton provides a solid platform for the chassis components. During two days of aggressive driving over a variety of road surfaces, I never detected any cowl shake or steering-column wiggle.

As befits a classic sports car, the Viper sports fully independent suspension, with unequal-length upper and lower control arms at each corner. With the exception of the front lower control arms, these pieces are fabricated from tubular steel. Engineers discovered that the Dodge Dakota pickup truck lower control arms had just the geometry they were seeking and adapted them to the Viper. Two stabilizing toe links are used with the rear lower control arms. Low-pressure gas-charged Koni coil-over damper/spring assemblies and front and rear anti-roll bars round out the underpinnings. Jounce to rebound, the Viper suspension has a full 8 in. of travel, so rough roads don't upset the car's balance. Furthermore, roll, dive and squat are well-controlled. Steering is by power-assisted rack and pinion, also adapted from the Dakota pickup. It offers positive on-center feel; response is immediate and obedient.

Part and parcel of the Viper's animalistic aura is a 17-in. tire and aluminum wheel package, the fenders bulging as if straining to contain the tremendous mechanical strength that lies just beneath the surface. The rear tires, Michelin XGT P335/35ZR-17s, are more than a foot wide, larger than the Corvette ZR-1's and the same size as those of the Lamborghini Diablo. Claimed lateral acceleration for the Viper is 0.95g, which seems plausible to me.

Inboard lie monster brakes, 13.0-in. vented discs with Brembo calipers. No ABS is available. Team Viper worked

Back to the Future with Canvas Top and Side Curtains

The Viper shuns outside door handles, roll-up side windows and factory-installed air conditioning, but a portable canvas convertible top and side curtains are standard equipment. The top is supported by five steel bows running laterally, and the entire mechanism expands accordion-like to stretch and latch in place between the windshield header and roof structural bar. Initially, I doubted that my 6-ft. 2-in. frame would clear the bows of the installed top, but with the driver's seat moved fully rearward and the seatback reclined slightly (a driving position I was most comfortable with), I could fit three fingers between the top of my head and the roof bow.

And it was great to note that the top didn't flap at all on an exhilarating run up to 150 mph. In fact, the Viper's interior became a much more civilized place for high-speed touring with the top in place, conversation or radio listening now possible over 100 mph. More or less traditional side curtains (not in place for the 150-mph run) round out the "weatherproofing" package, tabs snapping into a pair of slots atop each door.

The "window" part of the side curtains is see-through, flexible vinyl, like that used for the rear window on many convertible tops. At the leading edge of each curtain is a flap that can be pushed open to gain access to the door release handle inside the cockpit. Very elemental. The whole shooting match, side curtains, canvas top and bows, folds up and, if done correctly, will occupy about half of the available space in the minuscule trunk. Nobody ever said the Viper was practical.—*Ron Sessions*

very hard to build a hands-on machine that matches or betters the performance of the legendary 427 Cobra. One of the goals foremost in the minds of Chrysler engineers was the Cobra's vaunted 0–100–0-mph time of 15.0 sec. The Viper guys claim their snake can do it in 14.5 sec.; that's right, standing start to 100 mph to dead stop in 14.5 sec. *Road & Track* Engineering Editor, Dennis Simanaitis, did a little calculating and if Chrysler's braking claims are true, the Viper is capable of a best-in-class 211-ft. stopping distance from 80 mph, matching the Porsche 911 Turbo.

But let's not kid ourselves. The Viper isn't about numbers. It's about unbridled emotion on wheels. It's about explosive locomotion and the power to blast to 100 or 150 mph at will without working up a sweat. It's about balance and a 50/50 weight distribution that lets a skilled driver arm-wrestle difficult corners, approach and dance on the edge of the laws of physics without computer intervention. With a deep well of torque, you can accelerate out of corners faster, and with the massive binders, it's possible to brake later.

The Viper is sending little ripples of excitement through the ranks of Motor City car nuts and MoPar fans alike. The sort of excitement not experienced since the days of Hemi Cudas, 440 Six-Pack Dodge Challengers and winged Charger Daytonas. It has rekindled passions for an all-conquering, brawny-engine, front-midships roadster, passions that have been smoldering since the last 427 Shelby Cobras. As Team Viper leader Roy Sjoberg put it, "Chrysler intended to build a legend." Chrysler President Bob Lutz summed it up this way, "Viper is not for everyone. This car is only for the enthusiast who wants a great driving car and nothing more."

Strapped into the passenger seat of a Viper alongside Cobra sire Carroll Shelby at Willow Springs raceway, I felt tinges of nostalgia as he eased the V-10 roadster onto the track. With Carroll blasting up through the gears, I mused about the curious twists and turns that have occurred since the last Cobra was built: the oil shocks of 1973–1974 and 1979; the rush to me-too front-drive cars; the business failures of numerous exotic-car entrepreneurs; the terribly bastardized Chrysler's TC by Maserati; the seeming lack of an adventurous spirit in the Motor City. With the advantage of 25 years of hindsight, I just had to ask Carroll if he ever dreamed a car as exciting as his 427 Cobra would be built again. He half-turned, smiled broadly and shook his head no as we entered a series of delicious sweepers. Sometimes when you're driving the Dodge Viper RT/10, non-verbal communication is all that's needed.

SPECIFICATIONS	
Curb weight	**est 3300 lb**
Wheelbase	96.2 in.
Track, f/r	59.6 in./60.6 in.
Length	**175.1 in.**
Width	**75.7 in.**
Height	**44.0 in.**
Fuel capacity	22.0 gal.

ENGINE & DRIVETRAIN	
Engine	ohv 2-valve **V-10**
Bore x stroke	101.6 x 98.6 mm/ 4.00 x 3.88 in.
Displacement	**7990 cc/488 cu in.**
Compression ratio	9.1:1
Horsepower (SAE)	**400 bhp @ 4600 rpm**
Torque	**450 lb-ft @ 3600 rpm**
Fuel injection	elect. port
Transmission	**6-speed manual**

CHASSIS & BODY	
Layout	**front engine/rear drive**
Brake system, f/r	**13.0-in. vented discs/ 13.0-in. vented discs**
Wheels	composite alloy; 17 x 10-in. f, 17 x 13-in. r
Tires	**P275/40ZR-17 f, P335/35ZR-17 r**
Steering type	**rack & pinion,** pwr assist
Suspension, f/r	**upper & lower A-arms,** coil springs, tube shocks, anti-roll bar/**upper & lower A-arms,** toe links, coil springs, tube shocks, anti-roll bar

PHOTO BY RON PERRY

Prominent under the Viper's front-hinged hood is a snake's nest of ram-tuned intake runners. Mild camshaft timing with minimal overlap allows the Viper to be a pussycat around town—or brutally fast when called upon. Projected list price is $50,000; red is sole hue for 1992, but black and yellow ensue in 1993.

PHOTOS BY JOHN LAMM

Dodge Viper RT/10

Scoop: We become the first on earth to measure the speed of the Viper's strike.

BY KEVIN SMITH

The windows don't roll up. Matter of fact, there are no windows. Or outside door handles. Not much protection against the weather, either—the "top" is simply a section of canvas that stretches from the windshield header to the "sport bar." There is no hard top.

But what there is in this Dodge Viper RT/10 is a ten-cylinder, 488-cubic-inch powerplant producing 400 horsepower and a *Car and Driver*–measured 13.2-second quarter-mile time that makes it quicker than the altogether outrageous Chevrolet Corvette ZR-1.

And with the wind ripping new configurations in your eyebrows and the engine in full honk, you're not going to give one whit about absent windows or door handles. Because this Viper is one of the most exciting rides since Ben Hur discovered the chariot.

That's the whole point of the Viper. It's intended to go fast, stop hard, hang onto corners, and give everyone in sight—driver, passenger, and bystanders—a thrill that will make their day.

The Viper's massive ten-cylinder engine is mated to a six-speed transmission. An arresting plastic body hides a stout, steel-tube frame. There's also an unequal-length control-arm suspension, giant disc brakes, and wider-than-wide tires and wheels. And the entire exotic

PHOTOGRAPHY BY RICK CASEMORE

package weighs in at 3450 pounds, ready to rumble.

For about $55,000, you too can hop inside. Getting in begins with an awkward reach inside and a tug of the too-conventional Chrysler-style door handle, and the door opens. (An aside: your government says car doors need locks and would not grant the Viper an exemption, creating the curious circumstance of inside door locks that are readily accessible from the outside.) Lowering yourself into the cockpit is easier than dropping into a Corvette—which has taller rocker structures—but you must heed the label warning of door-sill heat from the enclosed side exhausts. Once settled in the supportive, leather-faced bucket seat, the view that greets you is at once unusual (white-faced instru-

ments in an unappealing gray panel material that looks and feels like double-ought sandpaper) and welcoming (familiar, straightforward controls and switches).

At first, you think the clutch feels oddly unyielding, but that's the brake; the pedals are offset to the left because of that great honking front-mounted motor squeezing its way into the passenger compartment.

Not even the most jaded long-time owner will ever reach for the Viper's ignition switch without a palpable twinge of excitement. Lighting off ten 799cc cylinders will always arouse the spirit. The sound that follows, unfortunately, doesn't. Good mechanical busyness and a hungry-mouth intake roar comes from the front, but the separate, five-cylinder, side-outlet exhaust plumbing gave the engineers fits. They couldn't tune in a melodious note *and* meet the federal 80-dBA noise limit. So the Viper sounds oddly like a UPS truck up to 3000 rpm, then it just roars like God's own Dustbuster.

Pulling out into city traffic for the first time is less intimidating than it might be in a car like this. The clutch action is moderately heavy but smooth in takeup, the shift lever moves with unexpected ease, and—noise aside—the engine doesn't seem to care if it's turning 1200 or 5000 rpm. It always offers great thrust and never bucks or spits or overheats. The steering effort is light, and the brake pedal has absolutely no lost motion. In short, despite its Hulk Hogan presence, the Viper is a pussycat to drive. Its standard six-speaker stereo even works acceptably, with just a bit of musical detail lost to three-digit wind speeds. One feature we miss is a good left-foot dead pedal.

We found the Viper oddly difficult to see out of—a strange complaint in an open car. The windshield frame is quite thick, close by the driver's eyes, and low on top. We were continually ducking down to spot

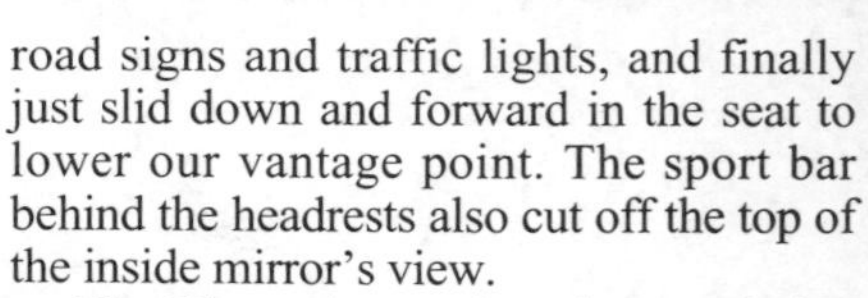

road signs and traffic lights, and finally just slid down and forward in the seat to lower our vantage point. The sport bar behind the headrests also cut off the top of the inside mirror's view.

The Viper tolerates the urban crush, but it lives and breathes on open roads. Be they fast or slow, it pounds along happily. It does prefer smooth surfaces, since its radically wide tires display the usual big-meat tendency to juke about trying to follow contours and ripples in the pavement. At elevated speeds and cornering loads, this action can be startling; a sharp bump in the middle of a high-g turn is bad enough if it causes momentary unweighting of the tires, but in the Viper that event can be accompanied by a sharp lateral feint. Even traveling straight, a big, suspension-pumping undulation in the road

can threaten to change heading.

That was the closest thing to a disconcerting trait we found. During our test drive on California Route 33's fast sweeping turns north of Ojai, the big car felt solid, secure, and predictable for the most part. It is nicely balanced, with a little polite understeer most of the time, and the Michelin XGT-Zs' breakaway is not particularly sudden. Aside from initial steering response that at first seems a bit abrupt, there is no trick to making this car do what you want. It even rides well, with minimal harshness and a sense of tremendous structural rigidity.

Which is not to pretend that no one will ever fall behind a Viper and be bitten. The engine makes so much power and the tires generate so much grip that the car can

work up tremendous speed without sweat or drama, and *that* could prove deceptive. High limits, once they're exceeded, mean big trouble.

Yet isn't that much of the appeal? Not the danger itself, but the awesome potential, the no-foolin' manner—those same qualities that demand respect also make a machine like this irresistible. It worked for the original Cobra, it still motivates lots of Cobra-replica buyers, and now it is firing enthusiasm for a wild new big-engined sports roadster from a highly unlikely source.

In the process—and this may be the great payoff—the landmark Dodge Viper RT/10 has brought a new word into the modern lexicon of automotive passion: "Chrysler." •

Vehicle type: front-engine, rear-wheel-drive, 2-passenger, 2-door roadster

Estimated price as tested: $54,640

Estimated price and option breakdown: base Dodge Viper RT/10 (includes $1700 gas-guzzler tax, $700 freight, and $2240 luxury tax), $54,640

Major standard accessories: power steering, tilt steering

Sound system: Chrysler/Alpine AM/FM-stereo radio/cassette, 6 speakers

ENGINE

Type ..... V-10, aluminum block and heads
Bore x stroke ..... 4.00 x 3.88 in, 101.6 x 98.6mm
Displacement ..... 488 cu in, 7990cc
Compression ratio 9.1:1
Engine-control system ..... Chrysler with port fuel injection
Emissions controls ..... 3-way catalytic converter, feedback fuel-air-ratio control
Valve gear ..... pushrods, hydraulic lifters
Power (SAE net) ..... 400 bhp @ 4600 rpm
Torque (SAE net) ..... 450 lb-ft @ 3600 rpm
Redline ..... 6000 rpm

DRIVETRAIN

Transmission 6-speed
Final-drive ratio ..... 3.07:1, limited slip

Gear	Ratio	Mph/1000 rpm	Max. test speed
I	2.66	9.2	55 mph (6000 rpm)
II	1.78	13.8	83 mph (6000 rpm)
III	1.30	18.9	113 mph (6000 rpm)
IV	1.00	24.5	147 mph (6000 rpm)
V	0.74	33.2	159 mph (4800 rpm)
VI	0.50	49.0	120 mph (2450 rpm)

DIMENSIONS AND CAPACITIES

Wheelbase ..... 96.2 in
Track, F/R ..... 59.6/60.8 in
Length ..... 175.1 in
Width ..... 75.7 in
Height ..... 44.0 in
Curb weight ..... 3450 lb
Weight distribution, F/R 50/50%
Fuel capacity ..... 22.0 gal
Oil capacity ..... 11.0 qt
Water capacity ..... 14.0 qt

CHASSIS/BODY

Type ..... steel tubing frame integral with body
Body material ..... fiberglass-reinforced plastic

INTERIOR

SAE volume, front seat ..... 48 cu ft
luggage space ..... 12 cu ft
Front seats ..... bucket
Restraint systems ..... door-mounted 3-point belts

SUSPENSION

F: ..... ind, unequal-length control arms, coil springs, anti-roll bar
R: ..... ind, unequal-length control arms with a toe-control link, coil springs, anti-roll bar

STEERING

Type ..... rack-and-pinion, power-assisted
Turns lock-to-lock 2.4
Turning circle curb-to-curb ..... 40.7 ft

BRAKES

F: ..... 13.0 x 1.3-in vented disc
R: ..... 13.0 x 0.9-in vented disc
Power assist ..... vacuum

WHEELS AND TIRES

Wheel size ..... F: 10.0 x 17 in, R: 13.0 x 17 in
Wheel type ..... cast aluminum hub, spun aluminum rim
Tires ..... Michelin XGT-Z, F: P275/40ZR-17, R: P335/35ZR-17
Test inflation pressures, F/R ..... 35/35 psi

CAR AND DRIVER TEST RESULTS

ACCELERATION Seconds

	Seconds
Zero to 30 mph	1.8
40 mph	2.5
50 mph	3.7
60 mph	4.6
70 mph	5.7
80 mph	7.6
90 mph	9.1
100 mph	11.7
110 mph	14.0
120 mph	16.6
Street start, 5–60 mph	4.9
Top-gear passing time, 30–50 mph	N.A.
50–70 mph	10.6
Standing 1/4-mile	13.2 sec @ 107 mph

Top speed ..... 159 mph

BRAKING

70–0 mph @ impending lockup ..... 193 ft
Modulation ..... poor fair **good** excellent
Fade ..... **none** moderate heavy
Front-rear balance ..... poor **fair** good

HANDLING

Roadholding, 300-ft-dia skidpad ..... 0.85 g
Understeer ..... **minimal** moderate excessive

PROJECTED FUEL ECONOMY

EPA city driving ..... 14 mpg
EPA highway driving ..... 19 mpg

Gestation of the Super Snake

Late 1987: Some "what if?" sketches of a big-engined sports roadster float around Chrysler's Advanced Design studio.

February 1988: Chatting in a hallway, Chrysler president Bob Lutz suggests that design v.p. Tom Gale have a go at a modern Cobra.

Spring 1988: A clay-model proposal for a concept car is completed. Lutz approves.

January 1989: The Viper concept car appears at the North American International Auto Show in Detroit. The public goes nuts, flooding Chrysler with "Build it, *please!*" mail.

March 1989: Chief engineer Roy Sjoberg starts to pick the 85 engineers who will become Team Viper; the team begins by studying low-volume vehicle development and reduced program lead times.

May 1989: Lamborghini Engineering is contracted to help adapt the upcoming T-300 pickup's 488-cubic-inch V-10 for sporting duty.

Fall 1989: A production-feasible body shape is finalized. Though much changed from the concept car, it looks the same.

December 1989: The first Viper chassis-development mule is on its wheels and running, using a V-8 engine.

February 1990: A Viper mule first runs under V-10 power.

May 1990: Chrysler chairman Lee Iacocca says "build it," allegedly after a fast blast down Highland Park's Oakland Avenue with Sjoberg in a prototype.

May 1991: Driven by Cobra creator Carroll Shelby, a prototype Viper paces the Indy 500. With practice and demo rides, the car logs 1800 miles, mostly at 130 mph.

November 1991: Auto writers are first turned loose on public roads and a racetrack in pre-production Vipers.

January 1992: First of the 200 or so 1992 Vipers are scheduled for delivery to customers; 2000 will follow in '93. After that, the market will get what it demands up to capacity of 5000-plus. The total investment of $50 million, by the way, has been small by Detroit standards: a fraction of what a routine face lift to a high-volume car costs. *—KS*

Technical Highlights

DAVID KIMBLE

A nine-pound sledgehammer is inelegant but effective, and you understand everything important about it the moment you see it. Same goes for the Dodge Viper RT/10. Its appearance—long hood, fat tires, open cockpit—promises brutal, bare-bones sports motoring, and in this case styling does *not* write checks the hardware can't cash.

The Viper's frame is welded rectangular-section steel tubing, with sturdy hoops fore and aft of the cockpit and massive rocker sections to make up for the lack of structure overhead. The result is stiff, but heavy. Without time to find where ounces could be pared, the engineers had to be safe and overbuild the frame. They missed their 3000-pound weight target by more than ten percent, and this is a lot of the reason.

"Modular full-face" wheels are something different, only partly because of their extreme rim widths (ten inches in front, thirteen in back). A cast outer section, called "the face," carries most of the load, due to pronounced negative offset. So the spun-aluminum inner section welded to it plays largely a tire-sealing role, and is thus made very light.

The ten-cylinder engine that motivates the Viper lost some 100 pounds when the block and heads were cast in aluminum alloy (with iron cylinder liners), but it's still big, long, and—at 716 pounds—quite heavy. For low hood height (and ease of manufacture on V-8 tooling, which ultimately turned out to be impractical), it uses a 90-degree V-angle rather than the 72 a V-10 would prefer. But interestingly, the throws on the forged crankshaft are arrayed as if the V-angle *were* 72 degrees. Although this gives a lopey, staggered firing sequence, it provides the most even firing possible without resorting to split crank throws.

There is modern microprocessor control for the ignition and sequential fuel injection, plus dual throttle bodies and separate intake plenums for the two five-cylinder banks. But the valvegear is old-school: a single camshaft down in the valley of the vee, driven by a double-row chain and actuating two valves per cylinder via hydraulic roller lifters, pushrods, and rocker arms. With 488 cubic inches and a ceiling of 6000 rpm, the complexities of more cams and valves seemed pointless. Output figures of 400 hp at 4600 rpm and 450 pound-feet of torque at 3600 rpm speak for themselves. *—KS*

Viper vs. Corvette ZR-1

There are no other cars quite like the Dodge Viper RT/10, but one obvious standard for comparison is Chevrolet's Corvette ZR-1, the reigning king of American speedsters and one of the fastest cars anywhere. Though open-car aerodynamics will keep the ZR-1's 170-plus-mph top speed safe, the Viper seems otherwise competitive. It weighs 69 pounds less than the last ZR-1 we tested (April 1991), and the big V-10 engine holds an advantage in both horsepower (400 at 4600 rpm versus 375 at 5800) and torque (450 pound-feet at 3600 rpm versus 370 at 5600). The Viper pulls even taller gearing—barely 1300 rpm at 65 mph in sixth and too tall to execute our 30-to-50-mph top-gear acceleration test—than the long-legged Corvette, with 3.07:1 final drive compared with the Corvette's 3.45. Ratios inside the two six-speed gearboxes are, coincidentally, almost identical.

Both transmissions offer the funky first-to-fourth "skip shift" feature in grudging deference to the EPA—though the Viper's operates in such a narrow speed range that it rarely intruded on our driving. Unlike the Corvette engineers, Team Viper did not have to avoid the gas-guzzler penalty. The ZR-1's 16/25 EPA mileage figures beat the Viper's 14/19 projections.

Massive seventeen-inch tires and wheels support both cars, but the Viper claims an edge in footprint with its 335/35 rear tires on thirteen-inch rims. The Corvette's are 315/35s on eleven-inchers. All else being equal (which it rarely is), more rubber and less weight should mean greater cornering grip. And the Viper's advantage in brake swept area—519 square inches to 422—should translate to superior braking. The Viper, however, does without the complexity of ABS—and of traction control, adjustable damping, and a driver's air bag, all standard on the ZR-1.

In the dollar derby, the Viper's relative simplicity seems to pay off. It carries a $54,640 estimated price tag, whereas the ZR-1 lists for $69,455. Early Viper buyers will likely pay a premium, however, while ZR-1s are currently being discounted.

Clandestine and admittedly incomplete instrumented testing of our Viper photo car lets us confirm some of Chrysler's performance claims and assess how the Viper stacks up against the ZR-1. We ran to a true 159 mph, with a smidgeon more to come. Chrysler lists a top speed of 165—probably in reach under ideal conditions. Our last ZR-1 went 171 mph. Advantage, Chevrolet.

This Viper's 0-to-60 time of 4.6 seconds backs up Chrysler's quoted 4.5 and edges the ZR-1's 4.9. In 0-to-100, however, aerodynamics and multivalve breathing start to tell, and the ZR-1 pulls ahead, 11.3 seconds to 11.7. The quarter-mile is almost even, the Viper posting a 13.2-second, 107-mph run, the ZR-1 13.4 seconds at 109 mph.

Chrysler touts the Viper's ability to do 0 to 100 to 0 (accelerate to 100 from a standing start then brake to a stop) in 14.5 seconds. We've never measured that "test," but our Viper needed all but 2.8 of those 14.5 seconds to reach 100, so we're skeptical, especially in light of a rather average 70-to-0 stopping distance: 193 feet, against the ZR-1's 159.

Was our pre-production Viper completely healthy, performing as sharply as final production examples will? Perhaps not. But, in any case, nuances of suspension geometry, balance, and control response can play almost as big a role in measured performance as the on-paper basics of mass, power, and footprint. We only had the chance to run the Viper on a wet skidpad, where it generated 0.85 g, so beating the ZR-1's 0.87 g is quite likely.

Clearly, the Viper RT/10 deserves to be mentioned along with the Corvette ZR-1 in any discussion of America's great performers. But the King doesn't look ready to relinquish his crown just yet. *—KS*

ROAD TEST

DODGE VIPER RT/10

FINALLY, THE LAST VIPER STORY YOU'LL HAVE TO READ (FOR A WHILE)

by Jeff Karr

PHOTOGRAPHY BY RICK GRAVES

By now, you're probably sick of the Dodge Viper RT/10. Never in modern times has a single vehicle received so much advance attention from the media. Dodge's redraft of the classic '60s muscle-car has been in the spotlight every inch of the way on the trip from showcar to showroom. There's likely not a single element of its creation that hasn't been dealt with in full detail. About a hundred times.

At *Motor Trend*, we've done at least our share of Viper-related coverage, and with good reason. For what it represents to the enthusiast driver and the future of the domestic auto industry, the Viper is news. Not because it points in a new or meaningful technical direction: It's underlying concept is and always has been an engineering

dead-end. No, the meaning behind the Viper lies in the spirit behind the car and in the method and swiftness with which it progressed from a wildly whimsical show car into an almost equally wild production car.

In a world where, as a rule, beautiful show cars mutate into bland parodies of themselves by the time they reach production, the Viper has made it to showrooms with nearly all its original venom intact. The result is a car that can upstage anything on wheels, wherever it goes. In the mold of the original Shelby Cobra, the Viper was to be a simple two-seater with a ton of power and only the most rudimentary creature comforts. The Viper makes the weather-tight Ferrari F-40 and Lamborghini Diablo seem positively sensible by comparison.

The hood is a challenge to open, but well worth the effort. The V-10 isn't really sophisticated in its technology, but it most certainly *is* in its power delivery. No excuses, no waiting; it's like having your right foot wired directly up to God's own adrenaline pump. While you're at it, you might want to ask His help in getting the hood shut again. You'll need it.

You'll want to replace the silly, door-mounted three-point belts with the dealer-option five-point harnesses before you even take delivery. First stop on the way home should be the nearest Thrifty Drug Store, where you'll need sunscreen (for obvious reasons), earplugs (buffeting in the cabin is substantial at speed), and a few cheap baseball caps (they blow off every time you hit redline in third).

The list of comfort and convenience features not found on the Viper is a long and complete one. Highlights of the not present and definitely unaccounted for include outside door latches, side glass, power mirrors, and air conditioning. There are also key technical features you can forget about, too, like multiple camshafts and four-valve heads, turbochargers, traction-control, anti-lock brakes, and an airbag.

Due, at least in part, to these notable omissions, the driving experience is unique among modern automobiles. The Viper's doors are opened by reaching inside the car and pulling the release handle. Locks are provided, but they're meant to satisfy some obscure DOT regulation and aren't expected to foil any save the dimmest of thieves. A key-chain-controlled alarm system provides a small measure of security and should give Viper owners the courage to leave sight of their car for short periods of time. In good neighborhoods. In daylight.

Once settled into the driver's seat, you'll find a comfortable but distinctly odd driving environment. The pedals are grouped far to the left to provide room for the engine. The effort levels of the brake and clutch pedals are firm, but not unpleasantly so. Nicely supportive leather-faced seats have a manual seatback angle adjuster and a squeeze-bulb type lumbar support adjuster. You also can set the steering wheel angle to suit your whims, so it's not too hard to come to terms with the car. Plainly visible, white-faced instruments tell you everything worth knowing in a classic, simple format.

The only remotely superfluous item in the interior is the six-speaker stereo system. It's reasonably good when the car is at rest, but wind turbulence and the exhaust drone nearly drown it out at high speeds. In terms of style points, the businesslike matte-finish dashboard comes across as clean and aggressive to some drivers, and simply cheap to others. The flimsy feel of the glovebox door met with disfavor, as did the fit of the door panels. The various interior bits are undergoing continuous refinement on the assembly line, according to Dodge; our test car was serial number 11, and there's still room for improvement.

Once underway, you'll notice plenty of heat in the footwells, on the coolant temperature gauge, and through the interior air vents. Don't worry about the engine; an auxiliary fan kicks in to keep things from boiling over. Do worry about your own composure, though. Heat is everywhere in the Viper. Watch the sidepipes and the smoking-hot (literally) doorsills when getting in or out. Learn to live with the 10 to 15 degrees of unwanted heat put into the vent system, or pay your dealer to install the optional air conditioner. Until you do, pray you never get caught in a summer thunderstorm. If you put on the cumbersome and leaky vinyl top and side curtains (which completely fill the trunk when stowed, by the way), you'll be trapped in a claustrophobic steam chamber that'll have you desperately scooping handfuls of cool air through the narrow zippered slits in the side curtains.

You wouldn't expect a roofless car to be any trouble to see out from, but the Team Viper people have defied physics this time out. The narrow-angle sideview mirrors are your best bet, but they have to be adjusted by hand—a task just as tedious today as it was back in the '60s with the original Cobra. The center mirror mostly shows you the B-pillar support bar behind the seats: Patrolmen will soon learn that a Viper is best approached from directly behind.

Particularly in freeway traffic, the Dodge is perennially at the center of a small traffic jam of onlookers, clotting main travel arteries like some sort of high-test coagulant. Busting loose of gangs of waving rubberneckers is no problem, however, given the V-10's daunting torque reserves.

Even at a mere 1200 rpm in sixth gear (about 60 mph), the Viper responds smartly to any application of throttle. That top cog only exists to squeeze a reasonable 22 highway mpg out of the car to please the feds; it's a safe bet people who've just dropped over 50 grand to buy The Most Useless Car In America aren't really concerned with fuel mileage. Better to keep the Dodge percolating along at about 2000 rpm or more, where the gratifying mount of thrust available is clearly worth the price of admission. Find a nice piece of snaking road, and you can just leave it in third for a whole tankful of gas. The Viper jumps off corners like nothing else; peaky, finicky turbos and little four-valve motors just can't equal the instantaneous and downright wonderful surge the Viper delivers.

And though the V-10 finds its roots in a future Chrysler truck engine, the help Lamborghini provided in its adaptation here makes it anything but a slow-witted foot soldier. The V-10 revs out to its 6000-rpm redline without running short of motivation. Ultimately, though, it's the Viper's overall flexibility that makes it thoroughly unique. There's power everywhere; you don't have to go looking for it, because it'll come looking for you.

The Viper's hard performance is always a crowd pleaser, but the audio/visceral presentation doesn't always play to rave reviews. The exhaust note at the impossibly low 400-rpm idle has the muffled, burbling cadence of a well-worn motorboat. At mid revs, there's a sonorous resonance between the two side pipes (each handling five cylinders) that has an aggressive edge, yet an "edited for airline use" sterility takes some of the sting out.

But at high engine speed, there's no debate, the Viper makes its best sounds. It's sort of an urgent, breathy hiss—the sound of a great volume of spent gasses spraying overboard at what sounds like incredible pressure. It's nothing like the uncorked bellow of a classic big-block Cobra, but in its way, the Viper's throaty hiss has a scary character all its own. It sounds like nothing you've ever heard, but some-

thing in your genetic makeup tells you its a dangerous sound. Be afraid.

Not many cars will outstrike the Viper. In our Feb. '92 issue, we recorded a ZR-1 Vette-beating 4.8-second trip from 0-60 in the Viper; that was in a prototype car at a place other than our normal test venue. For this article, at our normal track, the Viper used just 4.7 seconds to make the same trip; that's a tenth quicker than the number generated in a Callaway Twin Turbo for the August '91 issue, but not as quick as the best-ever ZR-1 time (4.3 seconds) we recorded a couple years ago. It stacks up well against the "brand-new" Shelby 427 S/C Cobra also tested in this issue, virtually mirroring its performance. Not bad for breathing through converters and real mufflers. The Viper could be quicker with more off-the line grip and quicker shift action from the otherwise agreeable Borg-Warner six-speed.

In a full quarter-mile run, the Viper stacks up about the same. It betters the ZR-1 by a small margin, but barely trails the Callaway: 13.0 seconds/112.7 mph versus 13.1/108.8 for the Viper. The Viper isn't the absolute quickest thing you can buy, but it certainly runs with a fast crowd.

As impressive as the Viper's performance under acceleration is, the chassis' handling numbers must be considered no less stirring. Our car made 0.97 g of lateral acceleration on the skidpad without great drama and hardly a scrap of body roll. The 50/50 weight distribution serves the Viper well, and the car is nearly neutral as it reaches the limits of its considerable grip. The cornering attitude is completely up to the driver, however. Get up to the limit, then lift off or step down sharply with your right foot, and the rear end will swing out into a glorious slide. Watch your steering closely though; the Viper lacks the catchable feel of a Vette or Nissan 300ZX when in heavy opposite-lock situations. The Viper's absolutely linear engine response will help a talented driver overcome the car's remote steering feel when playing on the skidpad, but not without some practice.

The same can be said of the Dodge's manners in the slalom. While the car's 68.4-mph average speed through the 600-foot course is good, the Viper isn't a car quickly mastered. You can bust loose its giant 335/35ZR17 rear Michelin XGT Zs at any point on the course. To drive the car well, new shadings of throttle control must be mastered; a ZR-1 is simple to drive in comparison.

The darker side of the Viper's nature is only truly apparent on the test track where you can explore big slip angles and arm-flailing oversteer. The car has an excellent chassis, extremely rigid and well suspended. But it's a classic Cobra-class muscle-car, and as such, requires high levels of driver skill and prescience if you wish to successfully venture over the limit. The Viper has no handy "eject" handle, no stabilizing ABS like most modern sports cars; misdirect the Viper's 400 horsepower or 3280 pounds at a bad moment, and you will be bitten. No car is foolproof, least of all the Viper.

On regular roads where smoother, saner tactics are the order, the Viper is a great treat to drive. The grip is positively tremendous, and, for the most part, shifting can be ignored. The steering, though hardly rich with information, at least has a solid, progressive feel, and the Viper turns in sharply and with hardly a hint of body roll. The wide, progressive Michelins are kept planted by the Viper's double wishbone suspension in most situations. Like other big-tired, firmly sprung cars, though, the Viper's composure is disrupted by genuinely rough or uneven pavement. The basic structure seems stout, much more so than most open cars, but when you really bound over bumpy surfaces, there are some secondary wiggles and squeaks in the bodywork, but no signs of protest from the tubular steel frame. The brakes are strong and fade resistant, but can be a little tough to modulate well at the point of lockup. Our best stop of 145 feet from 60 mph suggests the bias could still use a bit more tuning, or that ABS can't arrive too soon for future Vipers. There are many aspects of Cobra-class cars that are well worth reliving, but non-anti-lock brakes aren't among them.

The Viper is truth in advertising incarnate. Just as originally promised, the car's appeal is strictly visceral and emo-

With immense 17-inch Michelin XGT Zs front and rear, pizza-pan-size 13-inch vented disc brakes, and bulging fenders, the powerful Viper makes no secret of its intentions. In person, it's downright scary-looking.

tional. What the Viper offers is an 8-liter V-10 engine and just enough structure to lash one happy passenger on either side of its whirling driveshaft. To the people responsible for the Viper, nothing else really seemed necessary.

And it won't to the approximately 200 first-year Viper owners, either. All the stuff they're looking for is here, and not a bit more. The car's Spartan nature is at the heart of its appeal; some people will lapse into a happy adrenal spasm when they slide behind the Viper's wheel; others will wonder when they can park this breezy carnival ride and get into a real car.

Those complainers are the people who simply don't get it. Why you'd spend so much money to suffer such discomfort and inconvenience escapes them. If you've read this far, you probably know exactly why. Nothing you can drive packs the personality, makes the statement, or snaps your neck like the Dodge Viper RT/10. Nothing at all. **MT**

TECH DATA

Dodge Viper RT/10

GENERAL

Make and model	Dodge Viper RT/10
Manufacturer	Chrysler Corp., Detroit, Mich.
Location of final assembly plant	Detroit, Mich.
Body style	2-door, 2-passenger
Drivetrain layout	Front engine, rear drive
Base price	$50,000
Price as tested	$53,300
Options included	None
Ancillary charges	Gas-guzzler tax, $2600; destination, $700
Typical market competition	Chevrolet Corvette ZR-1, Acura NSX, Porsche 968

DIMENSIONS

Wheelbase, in./mm	96.2/2444
Track, f/r, in./mm	59.6/60.6/1514/1539
Length, in./mm	175.1/4448
Width, in./mm	75.7/1923
Height, in./mm	43.9/1115
Ground clearance, in./mm	5.1/130
Manufacturer's curb weight, lb	3280
Weight distribution, f/r, %	50/50
Cargo capacity, cu ft	11.8
Fuel capacity, gal	22.0
Weight/power ratio, lb/hp	8.2

ENGINE

Type	V-10, liquid cooled, cast aluminum block and heads
Bore x stroke, in./mm	4.00 x 3.88/ 101.6 x 98.6
Displacement, ci/cc	488/7990
Compression ratio	9.1:1
Valve gear	OHV, 2 valves/cylinder
Fuel/induction system	Multipoint EFI
Horsepower	
hp @ rpm, SAE net	400 @ 4600
Torque	
lb-ft @ rpm, SAE net	450 @ 3600
Horsepower/liter	50.1
Redline, rpm	6000
Recommended fuel	Unleaded premium

DRIVELINE

Transmission type	6-speed man.
Gear ratios	
(1st)	2.66:1
(2nd)	1.78:1
(3rd)	1.30:1
(4th)	1.00:1
(5th)	0.74:1
(6th)	0.50:1
Axle ratio	3.07:1
Final-drive ratio	1.54:1
Engine rpm,	
60 mph in top gear	1200

CHASSIS

Suspension	
Front	Upper and lower control arms, coil springs, anti-roll bar
Rear	Upper and lower control arms, coil springs, anti-roll bar
Steering	
Type	Rack and pinion, power assist
Ratio	16.7:1
Turns, lock to lock	2.4
Turning circle	40.7
Brakes	
Front, type/dia., in.	Vented discs/13.0
Rear, type/dia., in.	Vented discs/13.0
Anti-lock	Not offered
Wheels and tires	
Wheel size, in.	17 x 10.0/17 x 13.0
Wheel type/material	Cast aluminum
Tire size	275/40ZR17/335/35ZR17
Tire mfr. and model	Michelin XGT Z

INSTRUMENTATION

Instruments	180-mph speedo; 7000-rpm tach; oil pressure; fuel level; coolant temp; battery; digital clock
Warning lamps	Check engine

PERFORMANCE AND TEST DATA

Acceleration, sec	
0-30 mph	2.1
0-40 mph	2.8
0-50 mph	3.6
0-60 mph	4.7
0-70 mph	5.9
0-80 mph	7.7
Standing quarter mile	
sec @ mph	13.1 @ 108.8
Braking, ft	
60-0 mph	145
Handling	
Lateral acceleration, g	0.97
Speed through 600-ft slalom, mph	68.4
Speedometer error, mph	
Indicated	Actual
30	32
40	42
50	52
60	62
Interior noise, dBA	
Idling in neutral	70
Steady 60 mph in top gear	81

FUEL ECONOMY

EPA, mpg	13/22
Est. rangec city/hwy., miles	286/484

SECOND OPINION

Though more an entertainment device for affluent semi-adults than a car, the Viper is the most significant American automobile in decades. Yet its importance has nothing to do with its blazing acceleration, race-car handling, or styling that draws attention like Harrison Ford in the express line at Safeway. Instead, the Viper's distinction lies in how it was conceived, designed, and produced.

With the Viper, Chrysler tossed out Detroit's previous ponderous system, a practice that had (and has) as little hope of catching the Japanese as does a LeBaron of running down a Viper. With the Viper, Chrysler's top brass assembled its best and brightest, gave them an appropriate budget, and turned them loose.

The fruit of Team Viper beats a Ferrari 512TR in most performance parameters, is more fun to drive, *and* will attract more envious glances—at less than a third the cost. *—Mac DeMere*

COMPARISON
R&T
ROAD TEST

VECTOR VETTE VIPER

Attack of the Killer Vees

BY DOUGLAS KOTT
PHOTO BY GUY SPANGENBERG

For a moment think back to your childhood. Those formative years when Santa Claus wasn't make-believe but a living, breathing fat guy with an insatiable appetite for cookies and milk and questionable taste in clothing. Remember Christmas Eve, and the insomnia-wreaking anticipation of the things you'd find under the tree come morning? Visions of sugarplums dancing in your head? Heck no—train sets, Cox .049 U-control planes, slot cars and G.I. Joe in full combat fatigues; those were the essence of true sleeplessness.

Now we're older and a little more gray, and the list of things that takes away from our precious slumber is a much shorter one. Near the top is the prospect of spending the next day with more than 1400 horsepower divided up among three of the most exciting American supercars ever to grace asphalt. The Vector W8 TwinTurbo, perhaps the boldest, most powerful stylistic statement to ever roll on four wheels; Chevrolet's Corvette ZR-1, GM's technological flagship packing the punch of a 4-cam, 32-valve V-8; and the Dodge Viper, Chrysler's sensational, muscle-bound modern interpretation of the classic roadster.

Like kids on Christmas morning, we congregated at the entrance to our secret 1.0-mile road course, figuratively tore the wrapping off our presents, and spent the next 13½ hours putting America's supercars through their paces. Interspersed with track time were photo shoots (art directors can be so insistent) and forays to surrounding public roads to see how these highly tuned beasts behaved (or didn't) in traffic. After all the brake fluid had cooled, all the oil had gurgled back into the Vector's dry-sump catch tank and all the rubber dust floated to earth, the cars' strengths and weaknesses came to the fore. Distinct personalities emerged. With the aid of the 5th wheel, performance numbers were measured and recorded, and what numbers they are! So sit back and enjoy the fruits of American know-how; and hail to the tightly focused groups of designers, engineers and technicians that brought these cars into being.

Vector W8 TwinTurbo

"IT'S THE AUTOMOTIVE equivalent of the F-117A Stealth Fighter." Vector founder and president Gerald Wiegert must have repeated this statement to the point of hoarseness, but it's an excellent analogy. If Lockheed were in the business of building supercars with a distinct aircraft feel, the Vector W8 just might be the result. Under the engine cover lies not a Pratt & Whitney turbofan, but a 625-bhp 6.0-liter pushrod aluminum V-8, mounted transversely and force-fed with twin Garrett H3 turbochargers blowing through a pair of air-to-air intercoolers. The frame is of semimonocoque construction, with chrome-moly tubing forming the main safety cage and roll hoops. The aluminum monocoque structures, extending fore and aft of the safety cage, are bonded with epoxy and aircraft rivets.

Elsewhere, aircraft materials and technology abound, and every aspect of the car reminds you of its extremely specialized, hand-built nature. The body panels are formed from a combination of carbon fiber, Kevlar and unidirectional fiberglass, laboriously smoothed and filled, then painted to near perfection. Inside, in an interior entered through doors that swing up, Countach-style, is instrumentation that Vector terms an "aircraft-type menu-driven reconfigurable electroluminescent display"; we call it "fascinating, but hard to read quickly." All wiring and connectors on the car are to military specifications; all switches (with gold-plated contacts, of course) come from the same subcontractor who supplies the same pieces to General Dynamics and McDonnell Douglas.

The Vector is low—about belt-buckle height to me—and wide, a whopping 89.0 in. from mirror to mirror, just shy of the *wheelbase* of a Ford Festiva. You'll discover muscles you never knew you had when you enter the car. Sort of limbo-dance under the open door, set your butt down on a leather-covered sill maybe 8 in. wide, then lift your legs, pivot on your spinal axis and thread your legs into the footwell; then plop into the supportive leather-upholstered Recaro seat. The floor, covered with plush Wilton wool carpeting, is completely flat. There is massive front wheel arch intrusion, but you're sitting nearly in the center of the car, so the throttle and brake pedals are straight ahead. The view out front is through the largest, most steeply raked windshield in the business. The nose drops off so sharply that the immediacy of the pavement in front of you is startling.

Though leg room is generous, head room is tight for my 6-foot, 3-in. frame—when I straighten my back, my head presses firmly against the fixed-glass sunroof. Feature Editor Rich Homan, 6-foot-4 and taller of torso, just plain didn't fit. For those who do, the race-track fun begins when the turbo-muffled V-8 snarls to life, and the ratchet shifter, sunken into the sill to the driver's left, is moved into 1st gear.

■ ■ ■ ■

Utter ferocity arrives as the tach's moving-tape display skitters past 4000 rpm, the turbos hit full stride and what feels like the hand of God presses you back into the seat.

■ ■ ■ ■

The throttle is depressed, the throb of all that displacement fills the cockpit, and the car accelerates away impressively. Utter ferocity arrives as the tach's moving-tape display skitters past 4000 rpm, the turbos hit full stride and what feels like the hand of God presses you back into the seat. Box the shifter handle into 2nd, the electronically controlled wastegates exhale like Darth Vader with whooping cough, and the almighty process begins anew. At the 7000-rpm redline in 2nd, you're pushing 140 mph and trying to prevent that huge grin from squeezing your eyes shut. Third gear? We'd venture it's good for more than 200 mph, but Vector's powers-that-be haven't allowed top-speed testing.

On our sinuous little race track, the Vector felt a bit like a Kentucky Derby contender spinning tight circles in the corral. With more favorable gearing and more usable low-end torque at their disposal, the Vette and Viper would eat it up coming out of corners. The W8's handling composure was extremely stable and predictable, though, with De Dion suspension and big Michelin 315/40ZR-16 tires keeping the rear end firmly planted. Perhaps with more familiarity and left-foot braking to keep the boost up, the difference wouldn't seem so large. The Vector would be truly phenomenal at Willow Springs raceway, where its higher-speed acceleration potential could be safely and enthusiastically tapped.

No two ways about it, the Vector is intimidating in traffic—intimidating to other drivers who see its sharkish nose rushing up in their rearview mirrors, intimidating to those who drive it for the first time. Over-the-shoulder views out the back are an impossibility; you must rely on the two small but well-placed outside mirrors housed in somewhat droopy-looking fairings. "The expression 'What is behind me is not important' is given new meaning with this car," said one of us who obviously used the W8's forward thrust to his advantage. Adding to first-timers' traffic phobia are the extreme width, the near-central seating position, unusual shifter location and windshield pillars that are

■ From its ratchet shifter recessed in the doorsill to its aircraft-type switches, circuit breakers and instruments, the Vector's interior is an exotic, intriguing place to be. The 6.0-liter twin-turbo V-8 mounts transversely; air-to-air intercoolers can be seen on the right.

downright imposing. David Kostka, Vector's chief engineer and accomplished W8 jockey, made it all seem so easy as we hurtled down a section of the San Diego freeway en route to a refueling stop. Glance in the mirror, grab a shift, jab the throttle, stab the brakes. Child's play.

All were impressed, however, with the W8's civility. The noise level was

PHOTO BY DEAN SIRACUSA

Vector Avtech WX-3: Pushing the envelope

Gosh darn it, just when you thought you had the only Vector on the block, another gleaming W8 TwinTurbo whizzes past your house, nearly sucking your mailbox into its vacuum. An unlikely scenario, what with only 14 W8s built to date, but it could happen in, say, the ritzier parts of Hollywood. Lament not, because the talented craftsmen and engineers at Vector Aeromotive Corporation have been laboring tirelessly on a new car, dubbed the Avtech WX-3, that'll put you on top again in the game of supercar one-upmanship.

To call the WX-3 completely new would be misleading, as it keeps the sidewinder V-8 engine and retains the basic chassis of the W8 with its double A-arm front

Aircraft tech meets V-8 urge.

quite tolerable, noticeably quieter than the Vector I drove more than a year ago (March 1991). Steering effort is light, and more caster in the front suspension gives some self-centering feel that was absent from the earlier car. The interior is swathed in leather, the carpet is plush, and there's decent luggage space when the capacities of front and rear compartments are used. The Infinity sound system has the power and clarity to knife through engine noise, and its 10-CD player/changer is mounted high in the dash. And air conditioning—an absolute necessity with the W8's small cable-driven windows-within-a-window—quickly issues a flood of cold air through aircraft-style eyeball vents.

To fully appreciate the W8—and a car costing $448,000 requires full appreciation—it must be seen while under construction, internals exposed. You'd see parts milled from solid blocks of aluminum, then plated, polished or anodized. The massive Alcon racing brake calipers and 13.0-in. vented discs. The neatness of welds, the precision of the riveting. The sewer-pipe-size De Dion tube and the javelin-length trailing links of the rear suspension. When a sum approaching a half-million dollars is laid out for a mere car, the buyer expects it to make a strong statement. The Vector W8 TwinTurbo fills that need, inside, outside, underneath or over the road.

Chevrolet Corvette ZR-1

"To me," said one editor, "this car epitomizes 'Beauty and the Beast.' The beauty, of course, is the ZR-1's exterior shape (not its interior, by any means), and the beast is its engine." When parked next to the visually vociferous Viper and Vector, the Vette looks a little tame, perhaps a bit ordinary. But by its lonesome, the well-proportioned, muscular shape, set off with huge 17-in. wheels and tires (275s in front, 315s in back), drew both covetous glances and gaping stares around Newport Beach and on the freeways of Southern California. That says a lot about a design that was conceived more than a decade ago; the mystique of America's sports car is alive and doing quite well.

Something about the presence of a ZR-1 Corvette stirs the competitive juices of people near you—at least that's the case in our neck of the woods. Like giant wheeled gnats, sports coupes and econoboxes seem to swirl around the Vette, either tailgating, veering alongside or blatantly cutting you off. Baiting, inciting, challenging. I wished for a Sentra-size fly swatter to thin the ranks of these pesky intruders.

Luckily, ZR-1 owners can tap into the ultimate bug bomb, with 32 valves, four chain-driven camshafts and eight cylinders arranged in a 90-degree vee, and reduce those hangers-on to mere spots in the rearview mirror. More commonly known as the LT5, this 5.7-liter V-8 is a joint production overseen by Chevrolet, designed and developed by Lotus in Hethel, England, machined and assembled by Mercury Marine in Stillwater, Oklahoma, and shipped to the ZR-1 assembly line in Bowling Green, Kentucky. With the console-mounted power switch turned to Normal, there's a mere 210 bhp at your disposal. Twist it to Full, and the second of the two larger throttle bodies within the intake manifold comes into play, rousing the full wrath of 375 bhp. While there's ample torque just off idle, the LT5's fervor seems to increase exponentially up to redline at 7000 rpm, as it gulps air and rumbles the ground with its classic V-8 mechanical melody.

More so than the Viper or Vector, the ZR-1 demands driver involvement. Lots of arm is needed to twist the meaty, small-diameter steering wheel at either parking-lot speeds or 70 mph. Clutch effort tires left-leg calf and thigh muscles after a long urban drive. And the shifter to the 6-speed gearbox responds best to healthy tugs on its leather-wrapped knob. The harder you drive it, the better it seems to work. Putter around slowly, and the shifter's throws seem long and clunky; driven at low revs in 1st gear, the transmission sounds as if it's lubricated with crushed walnuts. And the worst thing? The dreaded CAGS (Computer Aided Gear Selection) solenoid, which can direct the shift lever from 1st to 4th under light throttle. Presumably a fuel-economy enhancer to avoid EPA's gas-guzzler label and stigma, CAGS encourages us to really stand on the gas in 1st to defeat it completely, effectively gulping more fuel. That's a loophole big enough to drive a Corvette through.

Entering the ZR-1 doesn't take as much doing as the Vector, but its high, narrow sills do slow down the act considerably. Once these hurdles are overcome, you fall into some of the finest sports-car seats ever made, with support and padding in all the right places. Driving position is quite good, aided by the electrically adjustable seats and manual tilt wheel. Though an immense transmission tunnel makes footwells narrow, there's a proper dead pedal and

and De Dion rear suspension systems and semi-monocoque construction. But its more rounded, elongated bodywork and its side windows that wrap around into the windshield give it a smoother, more contemporary look that retains the exciting essence of the original car. There are myriad interesting touches: the F40-style rear wing with servo-adjustable control surfaces for aero fine-tuning, triangular carbon-fiber housings for the rearview mirrors that can be retracted flush with the body for high-speed running, and functional ducts and vents of every description.

The intakes just aft of the WX-3's doors now feed air to twin coolant radiators that replace the W8's single front-mounted unit. The extra cooling capacity will be needed, because the now-7.0-liter V-8 will be making more power per unit displacement, thanks to new cylinder heads with four belt-driven camshafts and four valves per cylinder. For the first time, a normally aspirated version will be offered along with the twin-turbo model. Horsepower? Possibly 600–700 for the aspro, in excess of 1000 for the pressurized engine, say Vector spokesmen.

Chassis upgrades include lighter front suspension arms (beautifully crafted from elliptical-section tubing) and larger brakes (13.5-in. vented rotors, up from 13.0 in.) made possible by larger 18-in. wheels and tires. Just one car exists, doing double duty for car shows and development, with production slated for sometime in 1993. Price is expected to be—hold on to your hat—"in excess of $765,000." No one ever said keeping a step ahead of the Joneses was going to be cheap.—*Douglas Kott*

■ Despite its overly theatrical dash, the ZR-1's interior offers proper tools for the task of spirited driving—a grippy steering wheel, ideally positioned shifter, super-supportive seats (despite their leather covering) and well-spaced pedals. LT5 engine, too, wins great admiration.

easy access to clutch, brake and throttle. And there's enough head room and elbow space for even the tallest and most gangly of us.

If we had our druthers, we'd yank out the Vette's dash and instrument panel and replace it with something more readable and esthetically pleasing. As it stands, the panel is an odd mishmash of analog tach, oil-pressure, oil-temperature, coolant-temperature and volts gauges with a digital speedometer and a bar-graph fuel-level display. It's housed in a pod whose design can only be described as Gothic, complete with flying-buttress air vents. Also, the quality and fit of some of the interior's plastic pieces that are borderline in a $35,000 LT1 Vette have no place in the $65,000 ZR-1. On the plus side, controls for mirrors, lights, trip computer, radio and ventilation system are arranged roughly concentric with the steering wheel, within easy reach.

For the track, the routine is this: Buckle in, start the engine, then defeat the ASR (Acceleration Slip Regulation, an adjunct of the Bosch anti-lock braking system), turn the Power switch to FULL and twist the Selective Ride Control knob to PERF. Only then can you fully exploit the chassis, and there is plenty to exploit. Those massive Goodyear GS-Cs, working with suspension whose lightweight aluminum links look spindly by comparison, allow supreme confidence braking, turning into and accelerating out of corners. Mild understeer encroaches first, felt through the steering, which turns a little rubbery as the slip angles up front become more severe. Ease power on, and the tail can be made to step out, which happens with smoothness and with utter predictability. If lurid, power-on broadslides are your thing, this Vette will be your cohort for as long as the pavement (and tire budget) will allow.

Adroitness on the track thankfully doesn't translate to punishment on the street—with the shocks set in TOUR mode, the ZR-1's ride quality borders on cushy. In fact, most of us preferred the intermediate SPORT setting for everyday tooling around. Seats proved to be comfortable for the long haul, and gears we didn't even use on our road course—5th and 6th—drop engine noise to a mellow purr. In 6th, 75 mph works out to about 1900 rpm; great for listening to the ZR-1's Delco-Bose sound system, but poor for response, as even the mighty LT5 can't overcome an overdrive of 0.50:1. Drop down two gears and you can scoot around lagging traffic faster than you can say *Zahnradfabrik Friedrichshafen AG* (ZF for short, the manufacturer of the Corvette's gearbox).

In short, we're suckers for the ZR-1's siren song of acceleration, braking and overall handling competence. But Chevrolet needs to further differentiate the ZR-1 from the LT1 through a different interior treatment, specialized body panels, something. Much to the chagrin of current ZR-1 owners, the LT1 was given a ZR-1 rear bumper/taillight treatment, and its 5.7-liter pushrod V-8 now makes an even 300 bhp, edging much closer to the ZR-1's 375. Whether Chevrolet chooses to up the ante with the ZR-1 or lets it simmer while the LT1 is further fortified remains to be seen. For now, let's enjoy the ZR-1 for what it is—a world-class sports car that deserves all the respect and attention it continues to get.

Dodge Viper RT/10

LATE FOR THE 6 a.m. photo shoot, a staffer—who prefers to remain anonymous—was practically slaloming the Viper through sparse freeway traffic. Though closing speeds were not inconsiderable, more than one person being passed gave the Viper an enthusiastic thumbs-up sign. Were our man driving any other car, those hand gestures would have involved a single digit, but not the thumb . . .

The Viper, this Cobra reborn in the modern idiom, has that sort of effect on people. Bank robbers should use them as getaway cars; police and onlookers would gawk at the Viper and miss seeing the firearms, money bags and ski masks. Everywhere it's taken, the Viper attracts an unprecedented amount of attention, from everyone. And we're as susceptible as the next guy to the allure of its voluptuously formed bodywork, sassy fat tires and sidepipes. To Chrysler's credit, that allure extends far beneath the Viper's come-hither skin.

Under the resin-transfer-molded plastic panels you'll find the anatomy of a race car, with a space frame of welded tubular steel, unequal-length A-arm suspension all around and imposing 13.0-in. vented disc brakes without ABS. The rallying point of all this hardware is under the aggressively arched hood, and if engines were hand tools, the Viper's pushrod V-10 would be a broadax. It displaces 7990 cc or 488 cu. in., almost exactly a liter more than the thundering 427 of Cobra fame. Aside from the block and heads being made of aluminum, and the fancy interlaced runners and plenums of the induction

system, there's nothing technically extraordinary about it. It's plenty extraordinary in terms of output (400 bhp at 4600 rpm) and torque (450 lb.-ft. at 3600 rpm) and in how quickly it can move nearly 3500 lb. of Viper off the

■ ■ ■ ■

More than one person being passed gave the Viper an enthusiastic thumbs-up sign. Were our man driving any other car, those hand gestures would have involved a single digit, but not the thumb . . .

■ ■ ■ ■

mark and down the road.

"The biggest surprise is this engine," said one of us, "because I was expecting thundering, right-off-idle torque that would just make the car go in circles, but it's not like that at all. It has this fluffiness that's almost turbo-like." Indeed, the V-10 is docile off the line, but its torque is felt more immediately than with the ZR-1's LT5 engine or the turbo-softened low-end response of the Vector. And unlike many large-displacement pushrod engines of memory, this one blasts all the way to redline (6000 rpm) without a hint of breathlessness.

In anger, the Viper will reach 60 mph in 4.8 seconds and continue to storm up through the gears to a quarter-mile elapsed time of 13.1 sec., beating the Corvette but a second off the Vector's best. At play, the Viper is an easy car to drive around town, with very light clutch effort and compact, sturdy-feel-

■ Refreshing simplicity characterizes the Viper's cockpit, with its bold black-bezeled, white-faced gauges and only the most basic controls. Seats, though quite comfortable, could use firmer foam in their side bolsters. No complaints from the engine room, where the 8.0-liter V-10 delivers torque, torque and more torque.

ing throws to its Borg-Warner 6-speed gearbox (which also has a less objectionable version of the 1st–4th CAGS-type solenoid). The cockpit is so far back and the front wheels are so far forward that there's no wheel-arch intrusion whatsoever. So the pedals are to the far left of the footwell, crowded over by the enormous transmission tunnel to the right. While this arrangement feels perfectly natural after about 5 minutes of driving the car, it also doesn't leave anywhere to put your left foot when not using the clutch. There's no dead pedal, nor is there room for one.

The driver faces a very basic set of white-face, black-bezel instruments set into a light gray dashboard that has a sort of flocked texture. It adheres to the Viper team's basic design tenet of simplicity, and we like it very much. Seats have the basic manual adjustments of fore-aft positioning and a few degrees of seatback rake. Air conditioning? Not standard, but a system will be optional, installed at the dealer level. Side windows? None. There are side curtains and a canvas convertible top that store in the trunk, but they look hideous in place and are fussy to install. No matter, though. I'm sure most Viper owners will garage their cars on rainy days, for it is a car to be enjoyed with the wind ripping at your hair and the sun shining on your face.

It was sunny for our race-track session, and so were our dispositions after taking the Viper for a few laps. Its massive footprint—P275/40ZR-17 tires front, P335/35ZR-17s rear—gives amazing bite, a measured 0.96g on our skidpad. Body roll is virtually nil, and there is a sense of great chassis stiffness too, although the bodywork seems to jiggle around the chassis on some of the track's bumpier sections. Of the three cars, the Viper felt the most like a big, short-wheelbase go kart, with manners that were extremely predictable. But it's a car that will snap around much more quickly than the Corvette if power is fed too abruptly to the rear wheels while cornering. Treat it with respect, and the rewards can be enormous; get sloppy with the throttle, and all that torque will bite you.

On surface streets, we found the ride quality to be quite acceptable, more supple than its all-star track performance might suggest. Despite low ground clearance, the Viper handled seemingly treacherous driveways with surprising ease, thanks to its short overhangs and the steeply angled undersides of its bumper caps. Steering is fairly light and precise, though lacking the feedback of the Corvette's system.

PRICE	Chevrolet Corvette ZR-1	Dodge Viper RT/10	Vector W8 TwinTurbo
Base price	**$65,318**	**$50,000**	**$448,000**
Price as tested	**$68,960**	**$54,860**	**$489,800**
GENERAL DATA			
Curb weight	**3465 lb**	**3485 lb**	**est 3320 lb**
Test weight	3680 lb	3635 lb	est 3570 lb
Weight dist, f/r, %	52/48	49/51	45/55
Wheelbase	96.2 in.	96.2 in.	103.0 in.
Track, f/r	57.7 in./60.6 in.	59.6 in./60.6 in.	63.0 in./65.0 in.
Length	**178.5 in.**	**175.1 in.**	**172.0 in.**
Width	**73.1 in.**	**75.7 in.**	**82.0 in.**
Height	**46.3 in.**	**43.9 in.**	**42.5 in.**
ENGINE			
Type	dohc **V-8**	ohv **V-10**	twin-turbo ohv **V-8**
Displacement	350 cu in./5732 cc	488 cu in./7990 cc	365 cu in./5973 cc
Bore x stroke	3.90 x 3.66 in./ 99.0 x 93.0 mm	4.00 x 3.88 in./ 101.6 x 98.6 mm	4.00 x 3.63 in./ 101.6 x 92.1 mm
Compression ratio	11.0:1	9.1:1	8.0:1
Horsepower (SAE)	**375 bhp @ 5800 rpm**	**400 bhp @ 4600 rpm**	**625 bhp @ 5700 rpm**
Torque	**370 lb-ft @ 4800 rpm**	**450 lb-ft @ 3600 rpm**	**630 lb-ft @ 4900 rpm**
Maximum engine speed	7000 rpm	6000 rpm	7000 rpm
Fuel injection	elect. port	elect. port	elect. port
Fuel	prem unleaded, 91 pump oct	prem unleaded, 91 pump oct	prem unleaded, 91 pump oct
ACCOMMODATIONS			
Seating capacity	**2**	**2**	**2**
Head room	35.5 in	36.0 in.	35.5 in.
Front leg room	42.5 in.	44.5 in.	46.0 in.
Seat width	2 x 18.5 in.	2 x 18.5 in.	2 x 21.0 in.
Trunk space	11.6 cu ft	6.5 cu ft	7.0 + 4.0 cu ft
CHASSIS & BODY			
Layout	**front eng/rear drive**	**front eng/rear drive**	**mid eng/rear drive**
Body/frame	fiberglass/ skeletal steel	fiberglass + aluminum/ tubular steel	carbon fiber, Kevlar + fiberglass/tubular steel with aluminum panels
Brakes, f/r	**13.0-in. vented discs/ 12.0-in. vented discs;** vacuum assist, ABS	**13.0-in. vented discs/ 13.0-in. vented discs;** vacuum assist	**13.0-in. vented discs/ 13.0-in. vented discs**
Wheels	cast alloy; **17 x 9½ f, 17 x 11 r**	welded cast + spun alloy; **17 x 10½ f, 17 x 13 r**	modular alloy; **16 x 9½ f, 16 x 12 r**
Tires	Goodyear Eagle GS-C, **P275/40ZR-17 f, P315/35ZR-17 r**	Michelin XGT 2; **P275/40ZR-17 f, P335/35ZR-17 r**	Michelin XGT Plus; **255/45ZR-16 f, 315/40ZR-16 r**
Steering	**rack & pinion,** power assist	**rack & pinion,** power assist	**rack & pinion,** power assist
Overall ratio	15.6:1	16.7:1	16.5:1
Turning circle	40.0 ft	40.7 ft	46.2 ft
Suspension, f/r	**upper & lower A-arms,** transverse leaf spring, adj tube shocks, anti-roll bar/**multilink,** transverse leaf spring, adj tube shocks, anti-roll bar	**upper & lower A-arms,** coil springs, tube shocks, anti-roll bar/**upper & lower A-arms,** toe links, coil springs, tube shocks, anti-roll bar	**upper & lower A-arms,** coil springs, adj tube shocks, anti-roll bar/**De Dion tube,** upper & lower trailing links, diagonal link, coil springs, adj tube shocks, adj anti-roll bar
HANDLING			
Lateral accel (200-ft skidpad)	0.91g	0.96g	0.97g
Balance	mild understeer	mild understeer	mild understeer
Speed thru 700-ft slalom	63.6 mph	62.7 mph	60.6 mph
Balance	mild understeer	mild understeer	mild understeer

DRIVETRAIN

	Chevrolet Corvette ZR-1	Dodge Viper RT/10	Vector W8 TwinTurbo
Transmission	**6-sp manual**	**6-sp manual**	**3-sp automatic**
Gear: Ratio/Overall/(Rpm) Mph			
1st, :1	2.68/9.25/(7000) 58	2.66/8.17/(6000) 56	3.00/7.29/(7000) 70
2nd, :1	1.80/6.21/(7000) 86	1.78/5.46/(6000) 84	1.57/3.82/(7000) 137
3rd, :1	1.31/4.52/(7000) 118	1.30/3.99/(6000) 115	1.00/2.43/est (7000) 218
4th, :1	1.00/3.45/(7000) 154	1.00/3.07/(6000) 150	
5th, :1	0.75/2.59/(6050) 178	0.74/2.27/(4750) 160	
6th, :1	0.50/1.73/(4050) 178	0.50/1.54/est (3210) 160	
Final drive ratio	3.45:1	3.07:1	2.43:1
Engine rpm @ 60 mph in top gear	1360	1200	1955

ACCELERATION

	Chevrolet Corvette ZR-1	Dodge Viper RT/10	Vector W8 TwinTurbo
Time to speed, sec			
0–30 mph	2.5	2.0	1.9
0–40 mph	3.4	2.7	2.6
0–50 mph	4.2	3.6	3.3
0–60 mph	5.6	4.8	4.2
0–70 mph	7.0	5.8	5.1
0–80 mph	8.8	7.6	5.9
0–90 mph	10.8	9.2	7.0
0–100 mph	12.8	11.1	8.3
Time to distance			
0–1320 ft (¼ mi)	13.9 @ 105.0 mph	13.1 @ 109.0 mph	12.0 @ 124.0 mph

BRAKING

	Chevrolet Corvette ZR-1	Dodge Viper RT/10	Vector W8 TwinTurbo
Minimum stopping distance			
From 60 mph	142 ft	156 ft	145 ft
From 80 mph	256 ft	261 ft	250 ft
Control	excellent	very good	very good
Pedal effort for 0.5g stop	21 lb	22 lb	50 lb
Fade, effort after six 0.5g stops from 60 mph	24 lb	22 lb	50 lb
Brake feel	very good	average	good
Overall brake rating	very good	good	good

FUEL ECONOMY

	Chevrolet Corvette ZR-1	Dodge Viper RT/10	Vector W8 TwinTurbo
Normal driving	19.0 mpg	est 14.0 mpg	est 13.5 mpg
EPA city/highway	16/25 mpg	13/22 mpg	8/14 mpg
Fuel capacity	20.0 gal.	22.0 gal.	28.0 gal.

INTERIOR NOISE

	Chevrolet Corvette ZR-1	Dodge Viper RT/10	Vector W8 TwinTurbo
Idle in neutral	53 dBA	67 dBA	78 dBA
Maximum, 1st gear	82 dBA	87 dBA	90 dBA
Constant 70 mph	74 dBA	82 dBA	86 dBA

Subjective ratings consist of excellent, very good, good, average, poor.

Test Notes . . .

■ Our test ZR-1 was perhaps a lesson in car-to-car production variation: Compared with our original 1989 test car, this one was 0.7 sec slower to 60 mph, required 10 additional feet to stop from that speed, gripped around the skidpad at 0.91g down from the original's 0.94g, and was slightly slower through the slalom. But if you're looking for condemnation, you won't find it here. Even this comparatively slow ZR-1 ranks as an awesome sports car.

■ With the Viper, it's easy to be so enchanted with the personality of its huge engine that its fantastic handling can be overlooked. Okay, its 62.7-mph slalom speed is only very good, a consequence of its width and the driver's inability to see the cones beyond those big front tires. But lean on the Viper's considerable roll stiffness around the skidpad, and 0.96g—along with easy understeer and benign lift-throttle response—is the result.

■ Big, difficult to see out of, and filled with some of the oddest controls and instruments imaginable, the Vector W8 is a supercar that requires a lot of seat time to become accustomed to. Driven at part throttle, its big turbos take their time spooling up; moderate stopping power demands 50 lb. of leg effort (two to four times the norm). But hold the throttle fully open for a wink or two, and the Vector becomes a rocket.

But several things take the edge off the fun after a drive of, say, 10 minutes. The side pipes heat up the footwells in a hurry; in fact, they heat up the sills, the insides of the door panels and even the base of the windshield. Cockpit buffeting is pretty severe at freeway speeds, but removing the rear window that fits in the Viper's roll hoop lessens it considerably. And exhaust noise—what sounds like Titan booster-rocket thrust at full throttle—can make conversation and radio-listening lost arts at higher speeds.

Despite these criticisms, the Viper is the most exciting, invigorating, pulse-quickening car to show up on our doorstep in quite some time. Its arrival sent Senior Editor Joe Rusz reminiscing about a 1959 red Corvette he used to own. "My mouth got all dry because I couldn't wait to get in it and drive it around. And when I did, I had to show it to everyone I knew and give everyone a ride, and I had to swagger around and say, 'That's my car!' The Viper brought back that same feeling." I know Joe speaks for all of us at *Road & Track.*

Conclusion

WHAT A DAY: a roller-coaster ride of photo-shoot tedium offset by behind-the-wheel sessions in the quickest-accelerating, hardest-cornering, most provocatively styled cars on the continent. As the sun bids its hazy orange farewell over the foothills to the west, we're all in that strange state of exhaustion coupled with exaltation. We've sampled America's best supercars and come away mightily impressed with each.

The Vector shines with performance, aerospace technology, attention to detail and powerful, controversial styling. Daring to be different at the risk of non-acceptance, it's the high-tech Bohemian of supercars, and when the boost reaches its maximum of 10 psi, it's faster than a startled gypsy. The Corvette ZR-1, packing nearly 40 years of tradition, continues the bloodline with additional helpings of horsepower from the LT5 engine. It's the one car in our high-powered livery that we'd be comfortable driving day in, day out, rain or shine. And the Viper, O object of automotive lust, really sets us afire with its exquisite shape and theme of simplicity in all areas coupled with lots of torque-producing displacement. It puts Chrysler back on the supercar map in bold, block lettering.

It's just been one of those days, where the excitement of Christmas is in the air and you're justifiably proud to be an American.

Force Majeure

A QUARTET OF DODGE VIPERS IN A TAKE-NO-PRISONERS ASSAULT ON QUAINT EUROPEAN CHARM

by C. Van Tune

PHOTOGRAPHY BY THE AUTHOR

It's 11 p.m., and Tom Kowaleski looks concerned. One of his four Dodge Vipers is lost in the darkened wilds of Europe, far from our rendezvous hotel in Villeneuve, Switzerland. The driver, Team Viper General Manager Jean Mallebay-Vacqueur, is five hours late, and counting.

Jean was last spotted shortly after dawn, when the group departed from that Las Vegas of the Mediterranean known as Monte Carlo, now some 400 miles distant. The group (which includes Chrysler Public Relations and Marketing Manager Tom Kowaleski, fellow PR maven Tony Cervone, and a small contingent of Team Viper engineers) was to spend the day profiling their way to this point on placid Lake Geneva.

But something's gone awry. The boss is lost. And Kowaleski's not enjoying his meal.

With only seven humans to share piloting duties among the four Viper RT/10s Chrysler had shipped over for a preproduction shakedown run of BUX (build-up for export) models, it was necessary for someone to drive alone on the circuitous backroad route. Alone to read the indecipherable highway maps. Alone to follow the equally confusing highway signs. And alone to cross the heavily armed international borders. It made perfectly good sense: Having lived in France most of his life, Mallebay-Vacqueur would take the solo seat. If Jean became separated from the group, he'd have the best shot at linking up with us later at the hotel.

But by the following morning, the fourth Viper is still nowhere in sight.

Chrysler exports its wares to 38 countries, with about 1100 dealers in Europe alone. Jeeps, LeBarons, and minivans make up the bulk of the sales, but Viper soon will be among them. Europeans have been known to display a singleminded lust for fast, red sports cars, and this machine (which will carry the Chrysler, not Dodge, nameplate) is seen as a powerful invader from the land of giant engines, questionable quality control, and generous amounts of go-sideways torque. A V-10 Visigoth sent to slay the locals and set up retail outlets. And as the fourth-century Romans learned, it's better to be with the Goths than agin' 'em: Chrysler claims to have over 1000 export orders already on the books. However, no right-hand-drive models will be produced, and there are no plans to send Vipers to Japan.

A few modifications must be made before the car can be loaded onto transports at the New Mack assembly plant in Detroit and air-dropped into the 48th parallel. Headlamp patterns and side marker lights receive a minor retouching to conform to European laws, and the seatbelts will have their inertia reels moved from the doors to the center console—small tasks eagerly achieved to gain entry into such a huge marketplace.

One final necessary tweak results in the best modification that can be made to a Viper: piping the exhaust out the back of the car. Two black mufflers are recessed into the rear valance, and connect via a minor labyrinth of pipe to the catalytic converters that remain in their original locations near the leading edge of the side exhausts. Crude blockoff plates transform the original side-exhaust panels into non-functional styling gewgaws, which we feel should at once be removed and slam-dunked into the nearest recycling bin.

Even with its side-exhaust corpses in place (and your having to suffer the curbside embarrassment of explaining why something on this purpose-built car is there "for looks only"), the audible improvements of this setup far outweigh any negative side effects. The much lamented milk-truck exhaust tone magically is eliminated, because your ears now hear 10 cylinders at the rear of the car, not an uneven five on each side. Curious export Viper owners will find that removing the mufflers improves the growl and will help regain the 7 horsepower lost to this more restrictive system. Engine output has not been downrated from the advertised 400 horsepower at 4600 rpm, because dyno checks of production engines have shown a consistent 410 (or more) horsepower.

Our first stop on day two, sans the still-at-large Mallebay-Vacqueur, is for a fill-up on feloniously priced unleaded gasoline (either $7 or $12 per gallon, depending on whom of the mathematically illiterate among us you

Viper train fails to impress an elderly resident, who chases us away with a Regina Electrikbroom. Back on the autoroute, 170 mph is a big, windy thrill. En route to Paris, we pass everything but gas stations.

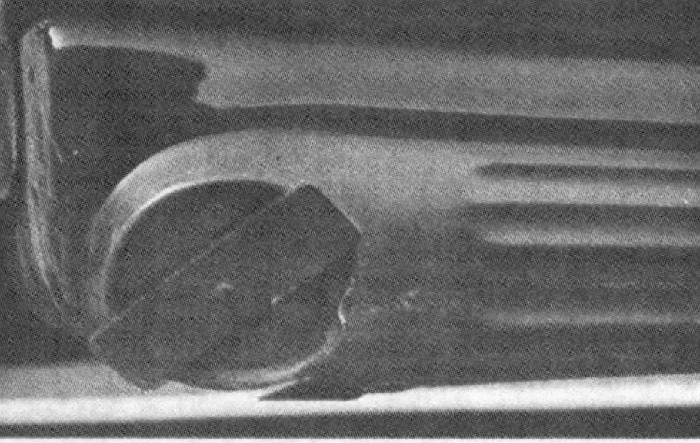

Before Chrysler can sell Vipers overseas, a few requirements must be met. Lighting laws demand modifications to the head and taillamps, safety regs force rearrangement of the seatbelts, and tight noise standards require the side exhausts be used for appearance only. Photos of prototype Viper show admittedly shoddy fitment. Production cars will have neater fascia trim and side pipe caps. Export models will be called Chrysler Vipers in every country but Germany, where an existing trademark pressures a renaming as Chrysler RT/10.

trust to figure the exchange rate/liter/gallon equation). As you know, Europe is a wonderful world of small-displacement engines and space-efficient cars. A Viper has neither of these qualities. All of which makes rampaging at high speeds through sleepy rural towns a thing of great precision and beauty. But don't consider this as willful, wanton endangerment by the drivers. They're being affected by a sort of speed blindness resulting from triple-digit running in an open car for the better part of the day. Ears are ringing, eyes are watering, and haircuts have been restructured into hideous knots of destruction. The aerodynamics of a Viper are such that no hat nor hairpiece known to mankind can withstand the vortices that whirl about the cockpit at just under supersonic speeds: High performance isn't good for your hair.

Wearing Michigan manufacturer plates, and therefore seemingly free from foreign jurisprudence, our Vipers nonetheless slow to the posted speed limit and chug through the little towns, then return to the act by accelerating at breakneck speed up to fifth gear. Even once achieving an indicated (slightly downhill) top speed of 170 mph. A Viper is in its element at these velocities, something not fully realized during our two test sessions in the States. The chassis is stable, precise, race-car tight, and the ultra-firm brake pedal feels perfect for rapid and controllable decels from our super-legal speeds. Even repeated hard use of the 13-inch rotors on France's busy autoroute can't produce any noticeable fade. All the Viper needs is ABS to rank as one of the best braking cars in the world.

But having nearly the best brakes on terra firma doesn't prevent us from being caught in a number of photo-radar traps along the way. Not that we're actually apprehended, but it's the sight of the five motorcycle-mounted gendarmerie pointing and staring at our finely honed nose-to-tail NASCAR drafting technique that gives us pause to reflect on our previous good fortune. Then we do exactly what you'd do under the circumstances: Nail it to the floor to add some fast distance, and then get the hell off the highway.

It's 5:30 p.m. It's Friday. We're entering Paris. Three million Parisians are exiting. Since this is France, everyone ignores lane control. Sometimes the sidewalk serves just as well. Horns honk, whistles blow, lights change to red. Nobody pays one whit of attention. Bump-drafting becomes the only method to ensure the three Vipers stay connected, though the unbelievable onslaught of battle-scarred Citroën 2CVs, smoke-belching M-A-N diesel trucks, and suicidal Vespa scooter jocks threatens to dismember our parade at every turn.

We fight our way past the Eiffel Tower and along the Champs Elysées (home to the best Burger King in Europe) without incident, other than the mini-riots caused by the sight of our three high-fashion rides in this land of the criminally well dressed. Once a minute, we answer the same questions from yet another minion of the pulsating mob: "Chrysler Viper...fifty thousand dollars, American...about two sixty-five kph," we reply again and again. By the fourth traffic light, we're dizzy from the attention; our minds wander: "Top secret cars...four hundred kph...one million dollars... we're highly paid American actors and rock stars...don't touch."

Riding solo, Jean Mallebay-Vacqueur will have to battle this melange alone. Of course, speaking the language, knowing the roads, and having proper change will help his case considerably.

Bonne chance. **MT**

DODGE VIPER GTS COUPE

Spiritual kin to a Sixties' racer

BY JOHN LAMM
PHOTOS BY THE AUTHOR

CARROLL SHELBY'S 427 Cobra was the inspiration for the Dodge Viper, so it's only logical that the second Viper body style also borrows from Cobra history. The company makes no apologies for it, Chrysler's vice president of engineering, François Castaing, declaring, "Where the Viper follows the classic sports-car gospel, this time we looked at some of the great Grand Touring cars such as the Cobra Daytona Coupe and Ferrari GTO."

Thus Chrysler has created a Viper coupe, the GTS, the stunning show-stopper at the Los Angeles automobile show. This is a car the designer of the original Cobra Daytona, Peter Brock, figures would outsell the open Viper sports car. *If* Chrysler were to build the GTS.

It wouldn't be a particularly difficult project. The Viper coupe is based very closely on the production machine. For the GTS, Chrysler started with a stock chassis and graced it with an exciting coupe body drawn at Chrysler Design in Detroit. While the hood and door skins started as stock Viper pieces, the rest is new, beginning at the front end, which has a fresh low air dam . . . probably too low for production. Stock hood openings still vent engine heat, but there are a few added holes up front. At the nose is a NACA duct to route cool air into the induction system. Atop each fender are louvers to relieve air pressure from under the fenders, adding downforce. Based on stock parts, all these pieces would retrofit to a roadster Viper.

A la Abarth, the GTS's roof has double bubbles to add head room for helmeted drivers. The car's shape tapers rearward down a hinged, all-glass liftback to a tail with a tall spoiler.

Put all those pieces together and what you have is a wonderfully powerful-looking car, a machine with great presence and character. Chrysler Design Chief Tom Gale explains, "The design is somewhat analogous to a recipe. I see a little bit of 250GTO and I certainly see a heavy sprinkling of Peter Brock's Cobra Daytona coupe . . . at least that's what was in the back of our collective minds." At one point, Chrysler planned to call this car the Viper Daytona Coupe, but thought that might cause confusion with the completely different—and worlds apart—production Dodge Daytona.

Chrysler President Bob Lutz adds, "To me the Pete Brock Cobras were always fascinating. They were aerodynamic yet brutal-looking, they radiated raw power, they were beautiful, they were American, they were all the right stuff. That's what we tried to recapture."

They did.

Very important to the car's image is the dark metallic blue paint with two wide white stripes, the same basic treatment as the Daytona coupes, though the GTS blue is much darker . . . and with a typical beautiful show-car finish. (Younger readers may not remember a time when countries were assigned specific colors in international racing; white and blue were ours.) At each corner is a polished aluminum road wheel of a new design. Incidentally, the GTS has been in the wind tunnel to confirm its stability at all speeds and attitudes.

Inside, the instrument panel (painted flat black), steering wheel and controls carry over from the production Viper. New are the seats, roll-up windows, 5-point competition seat harnesses and a fire extinguisher. A large outside competition-style fuel filler allows gasoline to be dumped into a 35-gal. fuel cell in the car's tail.

Underneath the GTS lies a stock Viper drivetrain, the 8.0-liter V-10 with its 400 horsepower and 465 lb.-ft. of torque. You'll note the side pipes have been blanked off, and the coupe has the "export" exhaust system, which exits out the back. Also, the GTS is, as

Castaing says with a grin, "minus the silencers, so it's pretty noisy . . . the way it should be." The standard 6-speed manual gearbox stays, as do the 4-wheel independent suspension and non-ABS disc brakes.

At 177.8 in. long, the GTS is 2.7 in. longer than the sports car, and the coupe body adds 4.7 in. of height to 48.7. The roof also adds weight. Castaing: "If we were to use the same plastics as the sports car's, we'd add something like 250 pounds to make the coupe because there is more glass and things like that. A case can be made that we could change the technology of the skin and keep the car weight close to the Viper weight. This is something we'll have to resolve, but if we were to keep it simple, this car would be slower than the Viper, at least in cornering."

There is a program to shed weight from the production Viper, the open model's target being 3000 lb., and those changes would be integrated into the GTS.

Lutz, looking at the far end of the GTS's performance scale, tells us, "Obviously the performance, especially the top end, is going to be way, way beyond the roadster's because the GTS is aerodynamically a much better vehicle, so while we haven't done any calculations or tests, I would imagine 200 mph wouldn't be very far away."

Given the Cobra heritage of both Viper models, it's not far-fetched to visualize them on race tracks. Castaing, who saw the Daytonas compete at Le Mans when he was a young man, cautions against too much visualization. "There has been speculation and discussion on whether the Viper

■ Viper GTS Coupe's interior uses stock dash and steering wheel, but has its own new seats and roll-up windows. Below, open Viper contrasts with the GTS Coupe and its large spoiler.

should be a race car or not," the engineer admits. "Today, Chrysler and Dodge are not very anxious to organize a factory-sponsored Viper racing effort, because we feel the car has so much personality that racing it and adding spoilers and things like that will not bring much to the image. If people want to race the car, that's fine, but we are not behind it.

"Right now planning for racing cars like the GTS is kind of foggy because FISA is still floundering around about what it's going to do after killing Group C," adds Castaing, a one-time Renault Formula 1 team manager. "There's some discussion that Group C will be replaced by some kind of supercar racing. Obviously I'd love to see the racing flavor of the Sixties come back, because those cars were terrific. I am not sure, however, modern technology will keep this dream feasible, because there's so much of a continuous technology race with electronics and the like. I don't know how much is reasonable. But, who knows?"

For that matter, who knows if the Viper GTS will ever be produced. Castaing explains, "Some of us would like to make a car such as this, if we keep it simple and it's not a very complex offspring from the Viper. There are a few challenges: The coupe is equipped with drop side-window glass and that forced us to remove the retractors for the seatbelts that we have in the doors, but we are planning to do that anyway to adapt airbags to the Viper. So there are a few technical things to be resolved.

"We have other new ideas on display in Detroit [at the auto show] like the Plymouth Prowler street rod, and we don't know if we need to do more for Dodge right now—which the GTS would do—or do something to promote Chrysler or Plymouth. It may be, in the end, a tough choice of which to do first."

Then the engineer smiles again and adds, "Maybe, if there is popular demand, we'll do everything."

Bob Lutz adds, "My personal feeling is that there are going to be well-to-do sports-car *aficionados* who are going to want a Viper roadster and a coupe . . . I'm certainly in that category. While the Viper turns everybody on, at some point a lot of people who live in rainy areas start thinking about the practical aspects, whereas with the coupe they can enjoy it all the time."

We certainly would.

Shades of the Cobra Daytona Coupe—Peter Brock

STEVE VOITA

"It's a pretty hot-looking piece," Peter Brock exclaimed as we walked up to the Viper GTS display. Chrysler Design Chief Tom Gale had given Brock a sneak preview of the GTS sketches a few months before, at the Monterey Historic Automobile Races, but this was the first time Brock had seen the Viper coupe for real. And his opinion counts, because as a young man working for Carroll Shelby in the Sixties, Peter Brock shaped the Cobra Daytona coupe.

What did Brock think about Chrysler doing the GTS? "Great, I'm pleased that you're honoring the idea," Brock told Gale. "Now I think it's come off far more successfully as a coupe than it did as a roadster."

With the turntable stopped to display the GTS's profile, Brock thought a bit about the Viper coupe and went on, "It's spectacular. It's big. I think one difference between the old designs and the new ones is that the tires have changed the proportions of the car."

Brock continued, "I think the proportioning of the roof to the rest of the mass of the car makes the GTS look a little bit smaller, where the Viper tends to look overly large. By putting a roof on the Viper, I think it ties the thing together a little better. The windscreen design works well and rounds the Viper coupe off.

"What's really nice on this car," Brock added, "and mimics the Daytona is that the roofline has the high spot over the driver's head, where it's supposed to be. We didn't have the double bubble on ours, of course, but the GTS's high point is where it should be, with the windscreen line much lower than the highest part of the roof."

As for Chrysler using the Cobra's paint scheme, Brock added, "If a different color had been used, you probably wouldn't notice the roofline change as much. One thing you can do to really accentuate the section is to run bold lines over it, and here you really see the bubbles in the roof. If the car was just a solid color, the bubbles would disappear as on an RX-7, where they're almost gone.

"The GTS is a real spectacular-looking car," Brock said. "It's not quite as boy racer as the roadster, and I like that. I think things are a little more integrated and toned down on the coupe, and I like that too. I'm sure if Dodge decides to put the GTS on the market, they'll sell more of these than roadsters."*—John Lamm*

SECOND STRIKE

First, Dodge created the Viper—the Shelby Cobra for the Nineties. Now it has built the Viper co

KEN OSBURN

show car inspired by the Cobra Daytona competition model. What's next, an assault on Le Mans?

Auburn Hills, Michigan—

The message is the war paint. Bold white racing stripes draped over gleaming ocean-blue pearlescent bodywork, a combination straight out of the 1960s—the competition livery, in fact, of the Shelby American Cobra Daytona competition coupes that beat Ferrari for the 1965 World Manufacturers' Championship for Grand Touring cars. On most of today's automobiles, a paint scheme like this would look ludicrous, but on the Viper GTS coupe, Dodge's new one-off show car, it's perfect. Nothing could better

BY RICH CEPPOS

PHOTOGRAPHY BY MICHAEL GASPAR

express the visceral connection between the Viper coupe and the legendary racing Cobra of twenty-eight years ago.

Although the Viper GTS looks like it is descended directly from the Cobra Daytona coupe, that simple observation does not explain its true purpose. Is it a pre-production version of a new model we will soon be able to buy? Is it, like the Cobra Daytona, destined to be one of a handful of competition cars built exclusively for the factory to race? Or is Dodge sending it out on the auto-show circuit simply to tease us until we weep?

We asked Neil Walling, Chrysler's director of advanced and international design—the man in charge of this project. "I'd be lying if I said to you the Cobra Daytona coupe isn't part of the inspiration for the GTS. But this is *not* a production car. Right now we have no plans to produce this car at all."

The project was commissioned, he explains, "to test the market, much like we did with the original Viper show car—to see if there's interest out there. And we do want to keep people talking about the Viper. There are a lot of things

Genesis of a show car: The Cobra Daytona competition coupe, above, was piloted by such luminaries as Dan Gurney, Phil Hill, Bob Bondurant, and Maurice Trintignant. It inspired the Viper coupe drawings that showed up in a photo of Tom Gale in our June 1992 issue. Right, the highly detailed three-eighths-scale wind-tunnel model.

SHOP PHOTOS BY KEN OSBURN

Assembly took place off Chrysler property in a top-secret shop in a Detroit suburb. Every one of the Viper roadster's body panels was modified in some way; the coupe's doors contain roll-up windows (the roadster has no side glass).

you could do with the coupe.'' Walling is grinning now. ''You could put it into production, or you could build a dozen or so and campaign them. But before you do anything like that, you have to do the first one.''

The first one, explains Walling, ''came as much as anything from designers saying 'wouldn't it be neat if . . .' '' Walling recalls Chrysler design chief Tom Gale asking what a competition Viper coupe or roadster would look like, ''and after that a designer does a sketch, and more and more people say, 'Gee, that would be neat.' It's an important part of creativity that we try to foster in allowing our people to have fun with their work—allowing people to pursue things that intuitively seem exciting and interesting. If you try to format all the design work, then suddenly everything gets distilled down to a neutral shade of gray.''

There is more to the Viper GTS, however, than just the designers' gut-level instincts about what a car like this should look like: There is science. A detailed three-eighths-scale clay model was developed during the design process for extensive wind tunnel tests. ''We thought of the GTS in terms of what you'd want if you were running it in competition,'' says Walling.

The wind tunnel work resulted in modifications to the front air dam, the addition of a pair of serious-looking vents atop the front fenders to relieve pressure from within the fender wells, a NACA duct in the hood for more efficient airflow to the engine, and a ducktail rear spoiler that adds downforce. Walling says, ''We achieved neutral lift

front-to-rear,'' which is a vast improvement over the Viper roadster. The need to improve the aerodynamics of the original Cobra roadster, by the way, was the sole reason the Daytona coupe was built; the roadster suffered from too much aerodynamic drag to be competitive with Ferrari's GTOs on high-speed European road courses like Le Mans.

It seems odd that Dodge would invest time and money improving the aerodynamics of a show car it has no plans to build. ''Well, you do have to do your homework,'' says Walling. ''When you show a car to the public, you're committed. You don't want to show it and then, if it goes into production, have to make a lot of changes to it because certain things aren't feasible.'' We'll take that as a clear signal that Dodge hasn't ruled out production after all.

Exactly what a production Viper coupe would be like to live with is something we can only speculate about, even though we did climb behind the GTS's wheel and take it for a spin around the block. Our encounter came at the Chrysler Technology Center just after the car was finished, and scant hours before it was to roll onto a truck for the cross-country trek to the Los Angeles Auto Show.

Chrysler's design staff was understandably concerned about the well-being of its only Viper coupe, which took three-and-a-half months to hand-make (the paint job alone required sixty hours). One stone chip would have had half-a-dozen fabricators jumping from tenth-story windows, so we trundled around at low speed just to get a sense of the car.

Underneath the new GTS bodywork is stock Viper running gear. Compared with a Viper roadster, the only difference we found during our short ride—aside from the absence of wind in the cabin—was a wonderfully raw, Gatling-gun exhaust note that bellowed from the specially fabricated pipes that exit at the rear of the car. Finally, a Viper that sounded sufficiently bellicose—just like Cobras immemorial. Too bad Dodge can't make them all this way.

Whatever course the GTS project takes from here on, the man who designed the original Cobra Daytona coupe is flattered by the attention the special Viper focuses on his twenty-eight-year-old handiwork. Pete Brock, now in his fifties, came to Shelby American and penned the Daytona after a stint as a GM designer. The Cobra racing coupe was built directly from Brock's intuitive freehand sketches; no blueprints were ever drawn, nor were any wind tunnel tests ever conducted. "We did that project off in the corner because only a few of us at the shop really understood anything about high-speed European racetracks." Brock's design worked so well that the Cobra's top speed leapt from about 160 mph to more than 180. Six Daytona coupes were built in all.

"Tom Gale came by and showed me the initial drawings of the Viper coupe," says Brock, "and he asked if I minded them doing it that way, with the paint and all. Did I mind? I was honored. Just the fact that they would share what they were doing was extremely considerate of them."

And what does Mr. Brock think of the new coupe? He likes it. "I think the concept they've come up with on the coupe better reflects what the Cobra Daytona is than the Viper roadster reflects what the original Cobra is. The coupe is a lot more in keeping with the theme they were looking for."

We can only add that we hope the rest of the world finds ocean-blue Viper coupes with broad white stripes as captivating as we do. The power of public opinion made the Viper roadster a reality; it might just cause something magical to happen all over again.

The Viper coupe's 1960s race-car design theme is underscored by touches like a Le Mans flip-up fuel door and a full-size spare tire sitting exposed under the hatch glass. (The FIA rules for mid-1960s GT cars required the Cobra Daytonas to carry a spare in the same location.) The GTS's engine is a stock 400-bhp Viper V-10, which exhales through special low-restriction pipes that exit at the rear of the car.

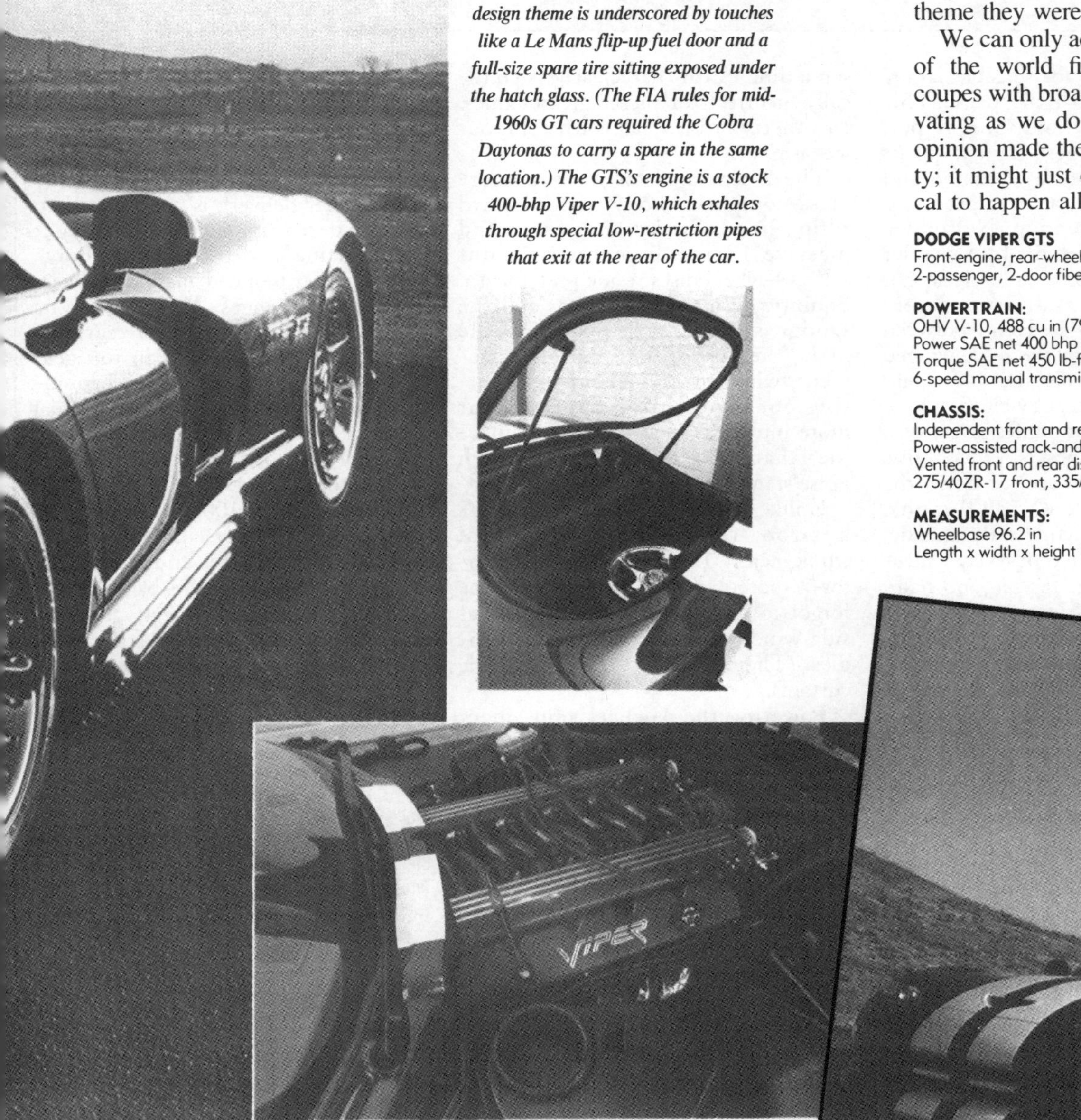

DODGE VIPER GTS
Front-engine, rear-wheel-drive coupe
2-passenger, 2-door fiberglass body

POWERTRAIN:
OHV V-10, 488 cu in (7997 cc)
Power SAE net 400 bhp @ 5200 rpm
Torque SAE net 450 lb-ft @ 3600 rpm
6-speed manual transmission

CHASSIS:
Independent front and rear suspension
Power-assisted rack-and-pinion steering
Vented front and rear disc brakes
275/40ZR-17 front, 335/35ZR-17 rear Michelin tires

MEASUREMENTS:
Wheelbase 96.2 in
Length x width x height 175.1 x 75.7 x 44.0 in

DODGE VIPER R/T10

PHOTO BY RON SESSIONS

If you're looking for sensible family transportation, turn the page. This isn't a car for the timid. Simply put, the Viper is the biggest-engine, quickest-accelerating, shortest-braking and grippiest-handling car ever produced by Chrysler Corporation. Nothing—not one machine—built by Chrysler during the heyday of the musclecar can touch the Viper in all-around performance. Not the 426 Hemi-powered Dodge Challenger or the high-winged Daytona. Forget about the 440-cu-in. Plymouth Barracuda, as well.

In fact, few automobiles short of a Shelby Cobra can best a Viper's time from 0-60 mph (4.8 sec) or down the 1/4-mile (13.1 sec @ 109.0 mph). Brutally fast and shamelessly archaic, Viper is the Sasquatch of today's automotive marketplace. It's big and hairy and muscular and flat scary to hang onto. A strange juxtaposition of high-fashion styling and pure, unadulterated force. A combination of shameless excess and back-to-basics logic at the same time. A car that, clearly, is politically incorrect for these "green" times. In other words, a car you'd sell your left arm for.

The 2-seat roadster wears a fiberglass-composite body with forward tilting hood. Sticky 17-in. tires and massive 13-in.-diameter (without ABS) 4-wheel discs reside underneath. Squinty, designer headlamps and the Dodge-signature "gunsight" grille make the appropriate muscle statement, while functional sidepipes jutting out beneath the doors are far more muted in tone than you'd imagine (thank the government's tough noise standards for that one).

Unlike Chevy's Corvette, Chrysler's hot-rod roadster offers a real trunk–nearly 12 cu ft.–and a legroom-for-7-footers interior, though someone forgot to include minor items such as side windows and exterior door handles. (There is a drop-in top and side curtains.)

Powering the 4-wheel amusement park is an all-aluminum V-10. At 8.0 liters (488 cu in.) of displacement, torque is its thing: 450 lb-ft @ 3600 rpm. Horsepower is a stout 400 @ 4600 rpm, and redline is a lofty 6000 rpm. Power delivery is smoother than you'd expect from such a heftyweight of an engine, and the standard Borg-Warner T56 6-speed manual delivers precise gear changes. Though, with its 0.50:1 ratio top gear, and rear-end cogs of 3.08:1, 100 mph in top gear equates to a mere 2000 rpm. Theoretical top speed is, thus, 300 mph. Actual terminal velocity is closer to 165 mph.

Production of the 1992 Viper was limited to just over 200 units; for 1993 it'll jump to over 1000, depending on the number of orders. Black joins red on the color palette, and there'll be other hues later on. At its base price of $50,000, and considering its perform-or-die attitude, Viper isn't a car for everyman/woman. If, however, you're looking for a V-10 Sasquatch of your own, hop in.

SPECIFICATIONS

Base price, base model	**$50,000**
Country of origin/assembly	**U.S.A.**
Body/seats	**conv/2**
Layout	**F/R**
Wheelbase	**96.2 in.**
Track, f/r	**59.6/60.6 in.**
Length	**175.1 in.**
Width	**75.7 in.**
Height	**43.9 in.**
Luggage capacity	**11.8 cu ft**
Curb weight	**3485 lb**

Fuel capacity	**22.0 gal.**
Fuel economy (EPA), city/highway	**13/22 mpg**
Base engine	**400-bhp, ohv, V-10**
Bore x stroke	**101.6 x 98.6 mm**
Displacement	**7990 cc**
Compression ratio	**9.1:1**
Horsepower, SAE net	**400 bhp @ 4600 rpm**
Torque	**450 lb-ft @ 3600 rpm**
Optional engine(s)	**none**
Transmission	**6M**
Suspension, f/r	**ind/ind**

Brakes, f/r	**disc/disc**
Tires	**P275/40ZR-17f, P335/35ZR-17r**
Steering type	**rack & pinion**
Turning circle	**40.7 ft**
Warranty, years/miles:	
Bumper-to-bumper	**1/12,000**[1]
Powertrain	**7/70,000**[1]
Rust-through	**7/100,000**
Passive restraint, driver's side	**door belt**
Front passenger's side	**door belt**

[1]combined 3/36,000 bumper-to-bumper and powertrain warranty optional at no extra cost